Sustainable Supply Chain Management

Sustainable Supply Chain Management: Industry-Driven Practices for a Resilient Future presents a groundbreaking exploration of how modern supply chains can strike a critical balance between efficiency, sustainability, and resilience in today's rapidly evolving global landscape. This comprehensive volume addresses the urgent need for supply chain transformation in the face of unprecedented challenges, including pandemics, climate change, and technological disruption.

Drawing from diverse global expertise spanning multiple continents and industries, the book seamlessly weaves together cutting-edge sustainability practices with practical industry applications. It provides actionable strategies for implementing green procurement, sustainable manufacturing, low-carbon logistics, and circular economy principles while maintaining a competitive advantage.

Designed for academics, scholars, practitioners, policymakers, and students, this essential resource demonstrates how organizations can build future-ready supply chains that create value for businesses, customers, society, and the environment through innovative, adaptive, and socially responsible practices.

Sustainable Industry Focused Supply Chain Management

Series Editors: Md. Mamun Habib, Erick C. Jones and Ferdoush Saleheen

This new book series is dedicated to providing a thorough analysis of the latest trends and practices in supply chain management across various industries. Topics covered include sustainable procurement and its operations, green logistics management, best practices in green manufacturing, reverse logistics, supply chain integration, industry-specific regulatory compliance and standards, collaborative partnerships, risk management, circular economy principles, low carbon logistics, transportation, performance metrics for sustainable reporting, sustainability standards and governance, and forecasting demand and supply fluctuations in this rapidly changing industry landscape. The series aims to connect theoretical concepts with real-life applications and operational challenges. Readers of the books in this new series will learn how to make decisions in uncertain business environments in today's fast-paced economy by implementing innovative strategies and adopting emerging trends in supply chain management.

The series will explore the contemporary challenges in supply chain management (SCM) such as global disruptions, sustainability issues, changes in international leadership and geopolitics, and their impacts on trade, business, and regulatory compliance. Readers will also gain insights into adopting best industry practices across various industrial sectors, including manufacturing and the service industry. The series will cover the challenges faced during recent disruptions such as the COVID-19 pandemic, post-pandemic recovery, and global unrest resulting from conflicts like the situation between Ukraine and Russia and discuss how organizations have maintained uninterrupted supply chain operations through their best-in-class strategies. Additionally, the series will analyze end-to-end risk management in supply chain operations, focusing on strategies to mitigate disruptions, maintain supplier relationships, and ensure smooth connectivity and operations. Included books in this series will explore the importance of synchronizing, adapting, and balancing global and local aspects of supply chain management to enhance efficiency and responsiveness. Furthermore, the series will speculate on the future of supply chain management, emphasizing the role of emerging technologies such as data analytics, artificial intelligence, and other innovations. Readers of the books in this series will gain valuable insights into the evolving landscape of SCM and its applications in the modern world.

Sustainable Supply Chain Management
Industry-Driven Practices for a Resilient Future
Md. Mamun Habib and Ferdoush Saleheen

For more information about this series, please visit: www.Routledge.com/Sustainable-Industry-Focused-Supply-Chain-Management/book-series/SIFSCM

Sustainable Supply Chain Management

Industry-Driven Practices for a Resilient Future

Edited by
Md. Mamun Habib
and
Ferdoush Saleheen

CRC Press
Taylor & Francis Group
Boca Raton London New York

CRC Press is an imprint of the
Taylor & Francis Group, an **informa** business

Designed cover image: Web Large Image (Public)

First edition published 2027
by CRC Press
2385 NW Executive Center Drive, Suite 320, Boca Raton FL 33431

and by CRC Press
4 Park Square, Milton Park, Abingdon, Oxon, OX14 4RN

CRC Press is an imprint of Taylor & Francis Group, LLC

ISBN: 978-1-041-29313-2 (hbk)
ISBN: 978-1-041-29536-5 (pbk)
ISBN: 978-1-003-77000-8 (ebk)

DOI: 10.1201/9781003770008

Typeset in Times
by SPi Technologies India Pvt Ltd (Straive)

Contents

Section III Technology and Digital Transformation

Section IV Circular Economy and Reverse Logistics

Section V Agricultural and Ecosystem-Based Value Chains

Preface

Global supply chains have become one of the defining infrastructures of the modern economy. They facilitate the movement of goods, resources, knowledge, and services across continents, connecting producers, manufacturers, logistics providers, and consumers within highly integrated networks. While these interconnected systems have significantly accelerated global economic growth, they have also introduced new vulnerabilities. Over the past decade, a series of disruptions, including pandemics, geopolitical tensions, climate-related events, and technological shifts, have exposed structural weaknesses in traditional supply chain systems and highlighted the need for more resilient and sustainable approaches to supply chain management.

As a result, supply chain management has entered a new strategic phase in which sustainability and resilience are increasingly recognized as fundamental priorities. *Sustainable Supply Chain Management* integrates economic performance with environmental stewardship and social responsibility, while resilience emphasizes the ability of supply chains to anticipate, adapt to, and recover from disruptions. Together, these perspectives redefine supply chains as adaptive systems capable of supporting both economic stability and responsible development.

This edited volume titled *Sustainable Supply Chain Management: Industry-Driven Practices for a Resilient Future*, under the series of *Sustainable Industry Focused Supply Chain Management*, brings together contributions from scholars and practitioners who examine how sustainability, governance, technological innovation, and ecosystem considerations are reshaping contemporary supply chain systems. The chapters collectively highlight how sustainable supply chains can function not only as environmentally responsible systems but also as strategic drivers of innovation, competitiveness, and long-term organizational resilience.

A distinctive feature of this volume is its emphasis on industry-driven practices. While theoretical frameworks provide important conceptual insights, the transformation toward sustainable supply chains ultimately occurs within real organizational contexts where firms must balance operational efficiency, stakeholder expectations, environmental responsibility, and technological change. By integrating conceptual research, systematic literature reviews, empirical studies, and applied industry perspectives, the contributions in this volume bridge the gap between academic scholarship and managerial practice.

This edited book is organized into five thematic sections consisting of 11 chapters that together offer a comprehensive perspective on sustainable supply chain transformation.

Section I – Performance Measurement focuses on the role of performance measurement frameworks in advancing sustainable supply chain strategies. The three chapters' contributions in this section explore how multidimensional performance indicators can capture operational efficiency, resilience, collaboration, and customer responsiveness across different levels of supply chain management.

Section II – CSR, Governance, and Social Marketing comprises two chapters and examines governance mechanisms and stakeholder engagement within sustainable supply chain systems. As supply chains operate across complex institutional environments, corporate social responsibility initiatives and stakeholder-driven governance structures increasingly shape responsible supply chain practices.

Section III – Technology and Digital Transformation represents two chapters and explores how emerging technologies are transforming supply chain operations. Advances in digital technologies, including artificial intelligence, blockchain, big data analytics, and the Internet of Things, are improving supply chain visibility, transparency, coordination, and data-driven decision-making.

Section IV – Circular Economy and Reverse Logistics includes two chapters and addresses the integration of circular economy principles into supply chain management. The contributions in this section highlight how reverse logistics systems, green manufacturing practices, and closed-loop production models can improve resource efficiency and environmental sustainability.

Section V – Agricultural and Ecosystem-Based Value Chains comprises two chapters and presents applied research examining sustainability challenges within agricultural and ecosystem-based supply chains. These studies illustrate how sustainable value-chain strategies can support rural livelihoods, environmental conservation, and inclusive economic development.

Taken together, the sections in this volume demonstrate that sustainable supply chain transformation requires an integrated approach combining performance measurement, governance frameworks, digital innovation, circular economy practices, and ecosystem-based value chain development. As global supply chains continue to evolve, organizations must move beyond efficiency-driven models toward adaptive systems capable of addressing economic, environmental, and social challenges simultaneously.

This book is intended for academics, scholars, policymakers, industry practitioners, and graduate students interested in supply chain management, logistics, sustainability, and operations strategy. By combining interdisciplinary perspectives with industry-oriented insights, the volume seeks to contribute to the growing body of knowledge on sustainable supply chain management while offering practical guidance for designing resilient and responsible supply chain systems.

We extend our sincere appreciation to all contributing authors whose work made this volume possible. Their research reflects the growing global commitment to advancing sustainable supply chain practices. We also acknowledge the editorial guidance and support provided by the CRC Press team at Taylor & Francis Group throughout the preparation of this book.

It is our hope that this volume will contribute meaningfully to ongoing discussions on sustainable supply chain management and serve as a valuable resource for researchers and practitioners working to design supply chains that are efficient, resilient, and responsible.

Editors
Md. Mamun Habib
Independent University, Bangladesh (IUB)
Visiting Scientist, University of Texas at Arlington, USA

Ferdoush Saleheen
Sharjah Maritime Academy, UAE

About the Editors

Prof. Dr. Md. Mamun Habib is professor and head of the Management Department at the School of Business and Entrepreneurship (SBE), Independent University, Bangladesh (IUB). He is a visiting scientist at the Department of Industrial Engineering, University of Texas–Arlington, USA. He also serves as visiting professor at UNIRAZAK, Malaysia; UCSI University, Malaysia; Polytechnic University of the Philippines (PUP), Philippines; Beijing Wuzi University, China; and Universitas Ciputra, Indonesia. He is actively involved in PhD supervision both locally and internationally. Currently, he supervises eight PhD scholars globally, while 13 PhD scholars have already been awarded under his supervision. He has more than 23 years of experience in teaching, training, workshops, consultancy, and research. His profile has been featured in *Britishpedia's Successful People in Malaysia and Singapore* section by the British Publication House, UK.

Prof. Habib has published around 250 research works, including conference proceedings, journal articles, and book chapters/books. Among them, more than 80 articles are indexed in Web of Science and Scopus. He also serves as editor-in-chief and editorial board member of various international journals. He has delivered keynote speeches at more than 100 international conferences across different countries and has conducted over 250 webinars and workshops locally and internationally. He completed his PhD and MS in Computer and Engineering Management (CEM) at Assumption University, Thailand, with outstanding performance. His core research areas include supply chain management, operations management, and research methodology.

Prof. Habib is an active member of several professional organizations, including IEEE (senior member); IEOM, USA; chartered fellow (FCILT), CILT International (UK); VCARE Academy, Canada; and AICRO, Australia, among others. He is also involved in business school accreditation, serving as AACSB Virtual Ambassador and ACBSP Evaluator, as well as the principal contact person for AACSB and ACBSP at IUB. In addition, he contributes as an academic participant in the QS World University Rankings and Times Higher Education Rankings. He is also connected with industry as a consultant with organizations such as AmCham (American Chamber of Commerce in Bangladesh), USAID, and BAFFA, among others.

Dr. Ferdoush Saleheen, PhD, FCILT, is an academic and industry leader in supply chain management, maritime logistics, and business transformation, with over 20 years of experience across multinational corporations and higher education institutions in Asia and the Middle East. He currently serves as head of the Maritime Logistics and Supply Chain Management Department at Sharjah Maritime Academy (SMA), UAE, where he contributes to research, teaching, and industry collaboration in global maritime logistics and supply chain systems.

Before joining SMA in March 2025 as an assistant professor, he served at the Higher Colleges of Technology (HCT), UAE, including as academic program chair. He has taught at BRAC University and the University of Liberal Arts Bangladesh (ULAB) and has delivered guest lectures at the University of Hull (UK).

Prior to academia, he held senior leadership roles in the FMCG, electronics, retail, and agribusiness sectors, including chief supply chain officer at Partex Star Group and Akij Food and Beverage Limited, and head of supply chain at LG Butterfly Bangladesh, leading large-scale supply chain transformation initiatives. Dr. Saleheen has conducted 100+ corporate training programs and authored 34+ publications, including Scopus/WoS journal articles, book chapters, and international conference papers. He serves on the editorial boards of several journals and is series co-editor of the *Sustainable Industry-Focused Supply Chain Management* book series with Taylor & Francis Group (CRC Press, USA).

He holds a PhD in Operations Management from Universiti Utara Malaysia, an MS in Logistics from the University of Texas at Arlington, and an MBA from Victoria University (Australia). He is chartered fellow (FCILT) of the Chartered Institute of Logistics and Transport (CILT International, UK). Dr. Ferdoush completed his PhD under the supervision of Prof. Dr. Md. Mamun Habib.

Contributors

Khaled Ben Abdallah
Higher Institute of Transport and Logistics and Laboratory of Research in Economy, Management and Quantitative Finance (LaREMFiQ),
University of Sousse
Sousse, Tunisia

Saleh Md. Arman
Department of Business Relationships and International Marketing
Poznan University of Economics and Business
Poznań, Poland

Prabha Arya
School of Management
GD Goenka University
Gurugram, India

Lawal Abideen Ayokunmi
Faculty of Technology Management and Business
Universiti Tun Hussein Onn Malaysia
Parit Raja, Malaysia

Benjamin Silas Bvepfepfe
School of Business
Sheffield Hallam University
Sheffield, UK

Sapna Dadwal
Department of Management Studies
DPG Institute of Technology and Management
Gurugram, India

Himanshu Gaur
Department of Management Studies
DPG Institute of Technology and Management
Gurugram, India

Vahid Ghomi
School of Business
Penn State Mont Alto
Mont Alto, Pennsylvania

Alica Grilec
Department of Logistics and Supply Chain Management
Higher Colleges of Technology
Fujairah, UAE

Md. Mamun Habib
Department of Management, School of Business and Entrepreneurship (SBE)
Independent University Bangladesh (IUB)
Dhaka, Bangladesh
and
Department of Industrial Engineering
The University of Texas at Arlington
Arlington, Texas

Suziana Hassan
Faculty of Economics and Management
Universiti Kebangsaan Malaysia
Bangi, Malaysia

Nabila Maruf
Dept of Economics, School of Business and Entrepreneurship (SBE)
Independent University Bangladesh
Dhaka, Bangladesh

Farnaz Ghazi Nezami
Department of Industrial and Manufacturing Engineering
Kettering University
Flint, Michigan

Banji Rildwan Olaleye
Business School
North-West University
Potchefstroom, South Africa

Fedy Ouni
Department of Maritime Logistics and Supply Chain Management
Sharjah Maritime Academy
Khor Fakkan, UAE

Mahour Parast
Ira A. Fulton Schools of Engineering
Arizona State University
Tempe, Arizona

Rauda Md. Ramli
Faculty of Economics and Management
Universiti Kebangsaan Malaysia
Bangi, Malaysia

Md. Sohel Rana
College of Business Administration
International University of Business Agriculture and Technology
Dhaka, Bangladesh

Umi Kartini Rashid
Johor Business School
Universiti Tun Hussein Onn Malaysia
Parit Raja, Malaysia

Firas Ben Romdhane
Higher Institute of Transport and Logistics
University of Sousse
Sousse, Tunisia

Mallika Roy
Department of Economics
University of Chittagong
Chittagong, Bangladesh
and
Central Queensland University
Rockhampton, Australia

Arsalan Safari
College of Business and Economics
Qatar University
Doha, Qatar

Md. Khaled Saifullah
Department of Economics, School of Business and Entrepreneurship
Independent University Bangladesh
Dhaka, Bangladesh

Ferdoush Saleheen
Department of Maritime Logistics and Supply Chain Management
Sharjah Maritime Academy
Khor Fakkan, UAE

Abul Hasnat Muhammed Salimullah
Department of Economics, School of Business and Entrepreneurship (SBE)
Independent University, Bangladesh (IUB) and University of Chittagong
Chittagong, Bangladesh

Noor Aslinda Abu Seman
Johor Business School
Universiti Tun Hussein Onn Malaysia
Parit Raja, Malaysia

Sonia Shahid
Athena Infonomics
Dhaka, Bangladesh

Section I

Performance Measurement

1 Agility and Resilience in Sustainable Supply Chain Management
Analysis of Enablers and Emerging Approaches

Vahid Ghomi, Farnaz Ghazi Nezami, Arsalan Safari, and Mahour Parast

1.1 INTRODUCTION

Environmental or global disruptions such as natural disasters, man-made disasters, worldwide outbreaks such as the COVID-19 pandemic, blockades, and political instabilities, as well as internal disruptions such as quality problems and operational failures, disrupt supply chain (SC) networks and the ability of firms to produce smoothly and deliver products and/or services to their customers. Such events have raised firms' awareness of the need to build a more resilient and sustainable system in all aspects of their operations, including their SC, to mitigate and respond to SC disruptions by becoming more resilient. This study analyzes the literature on supply chain resilience (SCR) and agility and then develops a framework to determine common concepts and best practices for achieving SCR.

SCR is about the ability of an organization to respond to disruptions and restore operations; SC agility is about the speed of a firm to quickly deploy resources and is an aspect of SCR that enables organizations to respond more effectively and efficiently to disruptions (Erol et al., 2024). SC disruptions can be categorized into supply, demand, process, environmental, and network/relationship disruptions (Gligor et al., 2019). To improve survivability and fulfill customer demands in competitive markets even during disruptions, firms need to be more resilient and agile to survive and achieve competitive advantages (Santoro et al., 2018).

1.1.1 Significance of Study and Research Questions

This chapter highlights the significance of firms' SCR and agility as two essential attributes of world-class SC management systems (Kochan and Nowicki, 2018). Several studies have investigated these concepts through conceptual model analyses

DOI: 10.1201/9781003770008-2

and their impacts on SC measures such as performance and sustainability (Ivanov, 2020; Nascimento et al., 2024). However, despite the plethora of studies on resilience and agility, no study to date has focused exclusively on the complex relationship between the two concepts. In this study, we provide a more comprehensive assessment of the enablers of these concepts by reviewing the literature in both domains. The proposed study answers the following research questions:

RQ1: What are the common enablers that integrate both concepts to avoid resource duplication in improving SC performance?
RQ2: What is a common framework that encompasses both SCR and agility?

The chapter is organized as follows. The second section presents the research methodology, and the third section addresses literature review studies on resilience and agility concepts at the firm level. The fourth section highlights the definitions and emerging antecedents/enablers/themes of these two concepts. The fifth section provides a synthesis of the findings of the study and introduces the house of resilience, and sections sixth and seventh present discussions and conclusions, respectively.

1.2 RESEARCH METHODOLOGY

Studies on SC agility and resilience can be traced back to Vastag et al. (1994) and Christopher and Peck (2004), respectively. Since then, a substantial amount of research has been undertaken using various methodological approaches. However, a review of the literature on agility and resilience reveals some inconsistency and ambiguity in concepts, whether they are synonyms, complementary, or distinct concepts. This ambiguity may hinder theoretical developments and consequently impose limitations on guidelines for real-world practices. Moreover, the literature reporting and analyzing these concepts in a systematic way remains scarce. This section aims to provide a holistic and interdisciplinary systematic literature review (SLR) on "Resilience" and "Agility" as two interrelated concepts in the SC, and identify the gaps and opportunities for further study and research. The scope of this analysis is at the firm level and is based on peer-reviewed articles in English, and we searched through Google Scholar and Web of Science for all SC agility- and residence-related articles published in 19 relevant top journals, using relevant keywords discussed below. Figure 1.1 shows the seven steps in the review process used to select articles.

Step 1 used publications in high-quality and interdisciplinary journals. Step 2 searched the terms "resilient", "resilience", "resiliency", "resili", "agility", and "agile". Step 3 sorted the journals based on the number of publications on SC agility and resilience. In Step 4, 2016 was determined as the cut-off point of the literature studies, as more than 50% of the total number of publications are after 2017. In Step 5, 62 SC agility and 157 SCR papers were selected, of which six articles were in both areas. Step 6 narrowed down the analysis to the most recent conceptual framework studies. Of the 225 articles after the second screening process, we determined there were 58 conceptual framework studies on SC agility and 87 on SCR. Of these, there were five articles in both areas. So, the total number was 140 articles. In Step 7, after scrutinizing all 140 studies, 71 completely relevant articles were selected (24 only in SC agility, 42 only in SCR, and 5 in both).

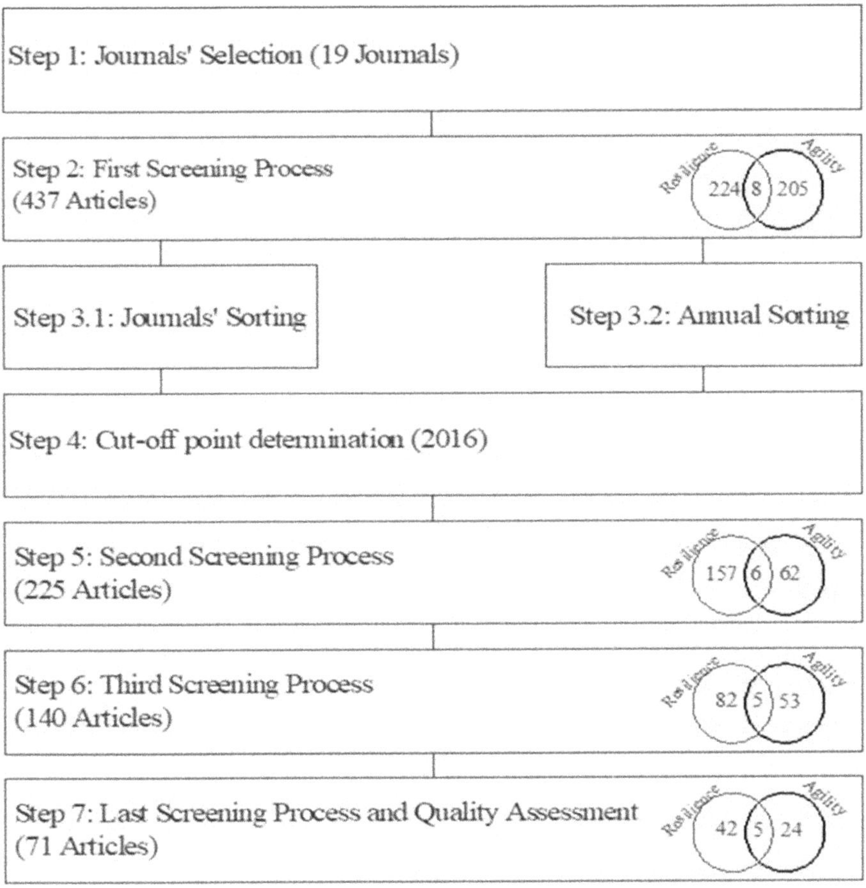

FIGURE 1.1 Methodology schema for selection of articles.

1.3 LITERATURE REVIEW

Bhamra et al. (2011) conducted a review on resilience in a broader view by focusing on resilience and its derivatives in medium-sized enterprises. They recommended further empirical research on organizational resilience, such as case studies and theory-building models. Tukamuhabwa et al. (2015) reviewed 91 articles and investigated the ripple effect framework to match the impact of digital technologies on different SC risks and disruption, and they asserted that firms should recover from SC disruptions to the same or better status with reasonable costs. Kamalahmadi and Parast (2016) discussed the pillars of SCR: SC reengineering, SC collaboration, SC agility, and SC risk management culture. However, they did not study the impact of various levels of disruption severity. Their study viewed agility as a component of SCR. Datta (2017) focused mainly on SCR in supply networks from an operations management perspective and identified interventions and mechanisms on a conceptual and theoretical level. Kochan and Nowicki (2018) discussed 16 SCR drivers, including agility, velocity, redundancy, efficiency, dispersion, organizational culture, anticipation, adaptability, and recovery. Ali and Gölgeci (2019) combined a set of drivers, barriers, theories, moderators, mediators, and research methods influencing SCR. To make a linkage between SCR and its enablers, they urged the need to enrich the literature with research exploring the boundary conditions (i.e., moderators) of SCR and mediating mechanisms between the known SCR drivers and consequences. Han et al. (2020) focused on the relation between resilience capabilities and performance. Finally, Queiroz et al. (2020) proposed a framework for operations and SC management during the COVID-19 listing adaptation, digitalization, preparedness, recovery, ripple effect, and sustainability. However, as a factor to check the effectiveness of a resilient system, they did not assume quality dimensions.

Sanchez and Nagi (2001) indicated that agility is a strategy for sustaining in an uncertain environment and responding to complexity. Upadhyay et al. (2022) reviewed 93 articles about lean and agile management in humanitarian SCs by analyzing the pre-disaster (mitigation and preparedness) versus post-disaster (response and recovery) factors. They concluded that all phases benefit from both lean and agility, and agility benefits the response phase the most. Al Humdan et al. (2020) listed the enablers of SC as proactive enablers and reactive enablers.

In summary, the SC literature has mainly focused on the SCR and agility themes separately. With the exception of the study by Gligor et al. (2019), previous studies did not consider the interrelationship of both concepts in a single framework. The study of Gligor et al. (2019) is also focused on the articles published before 2017. However, in this study, we focus on all relevant studies published after 2017, and discuss the themes of each concept on a detailed level to build a conceptual framework that consists of the main characteristics of both concepts.

1.4 AGILITY AND RESILIENCE THEME DEVELOPMENT AND DESCRIPTION

To address the research questions, we first implement a content analysis to determine the themes and enablers of each concept. Themes are used to define each concept before, during, or after disruptions occur.

1.4.1 Agility Themes

We determined 18 themes in the concept of agility, which are listed below.

Theme-A1: Responsiveness: When disruption occurs, responsiveness to rapidly changing markets and customers' requirements is a reactive and essential success factor for all firms. It is a dynamic capability of an agile SC to respond quickly to market changes and disruptions (Nath and Agrawal, 2020).

Theme-A2: Sense Changes: Sanchez and Nagi (2001) stated that firms in a challenging environment need organizational agility because it enables them to identify and react to unexpected disruptions. Managers need to distinguish between sensing (recognizing) and responding (reacting). Wamba and Akter (2019) expressed that SC agility is to recognize opportunities, realize threats, and transform SC processes, to align demand and supply, and improve adaptability and capability to respond quickly.

Theme-A3: Flexibility: The way firms collaborate with suppliers and customers to facilitate demand depends on the flexibility of their resources (Ghomi et al., 2021). Al Humdan et al. (2020) explained that a firm's competitiveness is maintained through flexibility and responsiveness.

Theme-A4: Adaptability: Cohen and Kouvelis (2020) state that adaptability and alignment are the foundations of an effective SC and lead to sustainable competitive advantage. Firms can improve their economic growth performance factors by investing in their core features such as ambidexterity, adaptability, and swiftness (Gunasekaran et al., 2019).

Theme-A5: Reconfiguration of Internal and External Operations: COVID-19 has caused companies to exercise SC reconfiguration for practical purposes. SC agility literature recognizes the role of reconfiguration in rapid sourcing of raw material, product design, development, testing, and distribution (Godwin-Geyi et al., 2020).

Theme-A6: Redundancy: Redundancy is defined as the ability to generate and maintain excess resources, including intangible cognitive, functional, and informative ones (Battisti et al., 2019). According to Altay et al. (2018), it is a dynamic and proactive capability to create backup capacities to improve the ability to respond to demand changes.

Theme-A7: Detection of Opportunities and Threats, Seizing Opportunities: Some studies defined SC agility as the capability to recognize the challenges and act swiftly by reconfiguring resources and processes, and also developing strategic partnerships and alliances (Dubey et al., 2020b, 2021b). Nath and Agrawal (2020) extended the idea and suggested that "sense" means the ability of an agile firm to search for opportunities and threats swiftly, and "seizing" means identifying potential market changes and tuning the current product portfolio or developing new products.

Theme-A8: Market Sensitivity: Christopher (2000) argued that an agile SC is market sensitive and capable of sensing and responding to actual demand. Al Humdan et al. (2020) recognize market sensitivity as one of the proactive enablers of SC agility, making the SC alert to capture market information. This capability helps in understanding the expectations of customers, while the lack of it could lead to failure (Godwin-Geyi et al., 2020).

Theme-A9: Mass Customization: Customer empowerment is the ability to provide customers with the tools and resources needed to be successful. As opposed to mass production, Gunasekaran et al. (2018) suggested that in an agile SC, mass customization is the capability to offer a wide range of product options in parallel at different markets. It also translated to a clear response in delivering superior customer values (Raji et al., 2021).

Theme-A10: Synchronizing Supply with Demand: Synchronization helps decrease the cycle time for new product development, which needs thorough planning to synchronize processes across SC, including internal and external partners (Brusset and Teller, 2017).

Theme-A11: Innovation: In an agile SC, any increase in the capability for adaptation can translate into improvisation. Dabić et al. (2021) concluded that the role of improvisation is to allow the development of competencies for adapting and moving quickly and smoothly, especially in unexpected contexts.

Theme-A12: Offering Transparent Added Value to Customers: Transparent customization seeks significant added value or lean product development more than cosmetic customization (Gunasekaran et al., 2018). Gunasekaran et al. (2019) emphasized the need for a major change using value-added processes for current products to survive in highly competitive markets.

Theme-A13: Strategic Partnerships and Collaboration: Firms in an agile environment need to extend their organizational boundaries through networking and consolidation of strategic partnerships. In such firms, collaboration is evidenced by the existing links of communication and information sharing between producers and buyers during disruptions (Raut et al., 2021). This also enhances flexibility and helps with resource planning (Dubey et al., 2021a).

Theme-A14: Visibility: SC visibility in an agile system improves the trust among the humanitarian stakeholders involved in disaster management (Dubey et al., 2020a). Visibility is essential to cooperation and agility and enhances trust, collaboration, and information sharing among workers during disruptions (Raji et al., 2021). Visibility also creates robustness and recovery capability against shocks. According to Wamba and Akter (2019), SC visibility helps SC partners to integrate, develop, and reconfigure resources. Gunasekaran et al. (2018) listed the Internet of Things (IoT), Industry 4.0, and blockchain technologies as three tools to enhance visibility in an agile SC. Thus, visibility emerges as another theme of agility.

Theme-A15: Employee Empowerment: This enabler emphasizes the importance of delegating authority and allowing employees to make independent decisions and implement them (Gligor et al., 2019). It refers to how firms provide their personnel with what they need to succeed (Gunasekaran et al., 2019).

Theme-A16: Shorter Delivery Lead Time: Agile systems provide better quality, shorter lead time, and thus higher customer satisfaction (Jajja et al., 2018). Gunasekaran et al. (2018) argued that shorter lead times improve competitiveness in a volatile market.

Theme-A17: Survivability: Survivability in an agile SC is the ability to continue to exist and provide society with vital products and services by adapting network design in rapidly changing environments (Raut et al., 2021). SC agility supports collaboration among SC partners, which increases the survivability of firms in a volatile market (Battisti et al., 2019).

Theme-A18: Social Media and Data Waves Preparedness: Nemkova (2017) argued that big data exist in niche global markets and social media. SC agility supports firms to quickly learn their customers' attitudes that have been successfully spread out in social media campaigns.

1.4.2 Resilience Themes

In this section, we present 18 themes for resilience.

Theme-R1: Return to Original State: This is the building block of a system to be recognized as resilient. The core characteristic of a resilient system is to bounce back to its original state after an upsetting event.

Theme-R2: Adaptability: A firm's adaptive capability refers to its capability to accept the ineluctability of transformation, adjust operations in response to certain eventualities, and utilize the required resources to address the new conditions during a disruption (Belhadi et al., 2021).

Theme-R3: Responsiveness: Responsiveness in a resilient SC is vital for rapid decision-making and actions after a disruption occurs (Kamalahmadi et al., 2021). Raut et al. (2021) suggested that firms need to focus on big data analytics to be highly market responsive.

Theme-R4: Resistance: Resistance against disruptions is the SC's ability to minimize the impact and duration of a disruption. It is the reactive ability of an SC to avoid a disruption (Polyviou et al., 2019).

Theme-R5: Business Continuity and Growth: To ensure continuity of operation, industries are required to manage operational challenges stemming from various types of disruptions, such as adverse weather, transportation network disruptions, equipment malfunctions, and incompetent service from logistics collaborators, to be sustainable.

Theme-R6: Knowledge Management: Firms need to spend time scanning the environment and learning to adapt and respond to changes swiftly. Belhadi et al. (2021) suggested that artificial intelligence (AI) and big data analytics are practical tools for initiating systems' adaptability by learning from the environment and making organized, highly reconfigurable systems.

Theme-R7: Mitigation of SC Variation: High flexibility allows for more significant internal and external variability in variety, volume, delivery, and supplier capabilities. SC variations create risk, propagate production errors, and can lead to a ripple effect in SCs (Li and Zobel, 2020).

Theme-R8: Flexibility: Ecological resilience recognizes the significance of flexibility in achieving a resilient SC. Some defined it as the magnitude of a disruption that a resilient system can tolerate and still persist in continuing its defined activities (Santoro et al., 2018).

Theme-R9: Long-Term Sustainability: Sustainability explores links between the environment, human and societal rights, and profitability. Ivanov (2020) linked an SC's performance to the resilience and sustainability of the SC. Raut et al. (2021) investigated the impact of big data analytics as a tool to connect and manage resilience and sustainability.

Theme-R10: Collaboration and Integration: Collaboration increases the preparedness level among SC members (Ali et al., 2018). Dubey et al. (2020b) suggested that firms use blockchain technology to enhance visibility and collaboration among SC members.

Theme-R11: Deployment of Dynamic Resources: Based on dynamic resource theory (Chowdhury and Quaddus, 2017), resilience emerges as a dynamic capacity that allows dynamic behavior of the outcomes. It hypothesizes that the deployment of dynamic resources will enable firms to better manage disruptions and therefore maintain higher performance.

Theme-R12: Identify and Anticipate Disruptions: Christopher and Peck (2004) stated that resilience is impacted by the capacity of customers and suppliers to identify and react to disruptions.

Theme-R13: Redundancy: Redundancy supports weak spots in an SC to prevent slowdowns or shutdowns. Increasing redundancy is critical for improving resilience (Polyviou et al., 2019). The redundancy can be interpreted as excess safety stock and production capacity, as well as diversified suppliers and multiple parallel modes of transportation.

Theme-R14: Survivability: Survivability is defined as the minimum levels of some financial or operational indicators that will allow a firm to maintain operations during a disruption and avoid bankruptcy (Ivanov and Dolgui, 2020). Survivability is the capacity to sustain and provide society with products and services for vital needs by adapting network designs and functions to radically changed environments (Ivanov, 2021).

Theme-R15: Visibility: It helps with traceability and refers to monitoring and tracking processes (Belhadi et al., 2021). There are tools to improve visibility, such as AI, IoT, and augmented reality technologies.

Theme-R16: Resource Reconfiguration: Reconfiguration means redirecting internal and external resources to tackle and adapt to a dynamic business environment (Ali and Gölgeci, 2019).

Theme-R17: Innovation: Innovation is the capacity to conceptualize a new product or idea (Belhadi et al., 2021). SC innovation capability factors consider internal skills, current competence and required competence sets, competence network, and optimal decisions (Raut et al., 2021).

Theme-R18: Contingency Planning: Contingency planning includes a wide range of actions to identify potential SC disruptions and minimize the effects of such disruptions. This is an efficient approach when the probability of an SC disruption is high and the disruption's expected impact is low.

1.5 HOUSE OF RESILIENCE

This section will synthesize the findings and our research questions. To answer RQ1, a summary of the findings is presented in Illustration 1, Figure 1.2, showing nine themes common to SC agility and resilience. From a managerial perspective, it is important to determine the typical characteristics of SC agility and resilience to avoid duplicating resources on the same themes and enhance the efficiency of each concept. In addition, by investing in the common themes, firms can enhance both concepts simultaneously. Through this summary, it can be seen that Responsiveness, Flexibility, Adaptability, Survivability, Innovation, Redundancy, Visibility, Reconfiguration, and Collaboration and Integration are the nine themes common to SC agility and SCR.

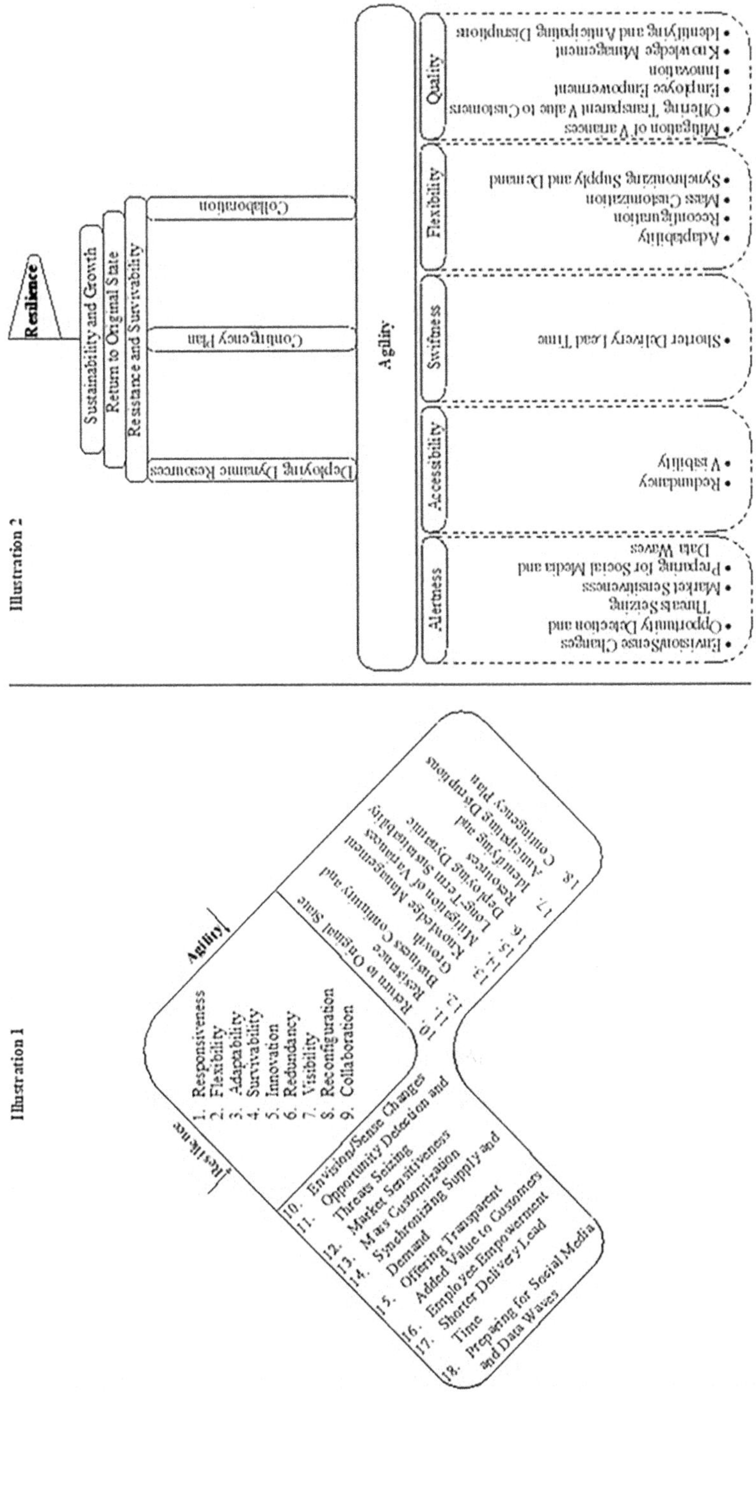

FIGURE 1.2 House of Supply Chain Resilience.

To address RQ2, we present our framework in the form of a "House of Supply Chain Resilience" (HSCR) that encompasses all the themes of agility and resilience (Illustration 2 in Figure 1.2). The HSCR summarizes the main components to achieve resilience, and we consider agility as its foundation. We use the construction analogy, and therefore, to reinforce the foundation's concrete, we assume each component of agility as a *Grade Beam* for more clarification. SC agility is a vital characteristic of a resilient SC (Ivanov, 2020). Al Humdan et al. (2020) discussed the main components of an agile SC: alertness, accessibility, swiftness, flexibility, and quality. *Alertness* in an SC is associated with opportunity-seeking tendencies during a disruption. Such companies sense changes to determine opportunities and threats (Dubey et al., 2020a), to respond to demand promptly. The second-grade beam to reinforce the SC agility is *Accessibility*. The third-grade beam is *Swiftness*, which is the ability to implement decisions quickly. The fourth-grade beam component is *Flexibility*. The final-grade beam is the overall *Quality* of firms' perspective toward agility.

Since the components of agility cover most of the themes of agility and resilience, they represent the foundation of the house of resilience to enhance the quality of responsiveness during disruptions. After we build the foundation, we need to complete the development of the HSCR by discussing the other three components of SCR: deployment of dynamic resources, collaboration, and contingency planning, which were explained earlier. The ceiling is the last stage to complete the *HSCR*. Therefore, we show three components for the ceiling: *resistance and survivability*, *return to the initial stage*, and *sustainability and growth after disruption*.

We assume resistance/survivability as the primary layer of the ceiling, since it is the central assumption of any resilient SC, and the primary dimension is the resistance against disruptions. However, to respond to disruptions, firms not only should reposition to their initial stages (the second layer of the ceiling), they should also strengthen their position in response to future disruptions, which are sustainability and growth (the third layer of the ceiling).

1.6 DISCUSSION

Considering the critical role of SCR and SC agility in improving organizational response to SC disruptions and environmental shocks, it is essential to empirically examine the impact of antecedents of resilience and agility on organizational performance and the extent to which different practices improve organizational response.

Given limited organizational resources, such an analysis provides practical insights for managers into allocating their (limited) resources to building capacity in areas with the most impact. One important research question in SCR is related to the cost-effectiveness of investment in SCR and agility. While investment in organizational capabilities to improve resilience is costly, organizations should look for cost-efficient approaches to make such investment financially viable. Thus, one immediate benefit of our proposed framework is the integration of different enablers into the resilient framework. There is a need for more research to explore the potential trade-off between investment in SCR capability and whether such investment can provide a long-term sustainable advantage.

1.7 CONCLUSION

Agility and resilience have appeared as intertwined concepts for a firm's survival and business sustainability. They are identified as necessary characteristics of a modern SC as tools to improve the firm's responsiveness to customers' demands during and after disruptions. Despite their growing popularity, there are still some ambiguities concerning the understanding of the concepts. Practitioners and academics often interchange these concepts due to the confusion and lack of consensus over the definition of SC agility and resilience. The first contribution of this work is to address a clear understanding of each concept: we identified enablers of firms' SCR and agility. Then, we determined nine themes common to SC agility and resilience. Learning about these enablers facilitates strategic planning for a sustainable development plan as it yields a strong foundation for organizations to hedge against economic loss. The second contribution is the development of the *HSCR*. Since most themes discussed in agility cover the themes of resilience, we show agility as the foundation of the house.

From a sustainability standpoint, a resilient SC considers success as a return to its original state, followed by sustainability and growth after disruptions. So, the strategic goal of a resilient SC should consist of short-term and long-term action plans for resistance and survivability, returning to the original state, and follow-up sustainability and growth. The proposed enablers help managers investigate an SC's weak and strong points, improve the SC's efficiency and effectiveness, and facilitate sustainable development planning. Also, identifying the common themes assists with designing and implementing more effective resource allocation strategies by avoiding repetitive work. While managers need to maximize agility and resilience in their SC, they must be mindful of the most efficient ways to achieve their overall SC goal and avoid overinvestment in similar capabilities. At the same time, investing in enablers to promote resilience and agility can improve a firm's position in sustainability, technology, and other progressive advancements that are essential to hold in the current competitive market.

One area of future research is evaluating contingency perspectives to explore how organizational and contextual factors influence resilience and agility. Examples include a comparative assessment between manufacturing and service SCs, and small or medium-sized enterprises versus large organizations. It would also be insightful to assess how resilience and agility capabilities should be developed based on the origin of the disruptions.

REFERENCES

Al Humdan, E., Shi, Y. and Behnia, M. (2020), "Supply chain agility: A systematic review of definitions, enablers and performance implications." *International Journal of Physical Distribution & Logistics Management*, Vol. 50, pp. 287–312.

Ali, I. and Gölgeci, I. (2019). "Where is supply chain resilience research heading? A systematic and co-occurrence analysis." *International Journal of Physical Distribution & Logistics Management*, Vol. 49, pp. 793–815.

Ali, I., Nagalingam, S. and Gurd, B. (2018), "A resilience model for cold chain logistics of perishable products." *The International Journal of Logistics Management*, Vol. 29, pp. 922–941.

Altay, N., Gunasekaran, A., Dubey, R. and Childe, S.J. (2018), "Agility and resilience as antecedents of supply chain performance under moderating effects of organizational culture within the humanitarian setting: A dynamic capability view." *Production Planning & Control*, Vol. 29, pp. 1158–1174.

Battisti, M., Beynon, M., Pickernell, D. and Deakins, D. (2019), "Surviving or thriving: The role of learning for the resilient performance of small firms." *Journal of Business Research*, Vol. 100, pp. 38–50.

Belhadi, A., Mani, V., Kamble, S., Khan, S.A.R. and Verma, S. (2021), "Artificial intelligence-driven innovation for enhancing supply chain resilience and performance under the effect of supply chain dynamism: An empirical investigation." *Annals of Operations Research*, Vol. 333, pp. 1–26.

Bhamra, R., Dani, S. and Burnard, K. (2011), "Resilience: The concept, a literature review and future directions." *International Journal of Production Research*, Vol. 49, pp. 5375–5393.

Brusset, X. and Teller, C. (2017), "Supply chain capabilities, risks, and resilience." *International Journal of Production Economics*, Vol. 184, pp. 59–68.

Chowdhury, M.M.H. and Quaddus, M. (2017), "Supply chain resilience: Conceptualization and scale development using dynamic capability theory." *International Journal of Production Economics*, Vol. 188, pp. 185–204.

Christopher, M. (2000), "The agile supply chain: competing in volatile markets." *Industrial Marketing Management*, Vol. 29, pp. 37–44.

Christopher, M. and Peck, H. (2004), "Building the resilient supply chain." *The International Journal of Logistics Management*, Vol. 15, pp. 1–13.

Cohen, M.A. and Kouvelis, P. (2020), "Revisit of AAA excellence of global value chains: Robustness, resilience, and realignment." *Production and Operations Management*, Vol. 30, pp. 633–643.

Dabić, M., Stojčić, N., Simić, M., Potocan, V., Slavković, M. and Nedelko, Z. (2021), "Intellectual agility and innovation in micro and small businesses: The mediating role of entrepreneurial leadership." *Journal of Business Research*, Vol. 123, pp. 683–695.

Datta, P. (2017), "Supply network resilience: A systematic literature review and future research." *The International Journal of Logistics Management*, Vol. 28, No. 4, pp. 1387–1424.

Dubey, R., Bryde, D.J., Foropon, C., Graham, G., Giannakis, M. and Mishra, D.B. (2020a), "Agility in humanitarian supply chain: An organizational information processing perspective and relational view." *Annals of Operations Research*, Vol. 319, pp. 1–21.

Dubey, R., Bryde, D.J., Foropon, C., Tiwari, M., Dwivedi, Y. and Schiffling, S. (2021a), "An investigation of information alignment and collaboration as complements to supply chain agility in humanitarian supply chain." *International Journal of Production Research*, Vol. 59, pp. 1586–1605.

Dubey, R., Gunasekaran, A., Bryde, D.J., Dwivedi, Y.K. and Papadopoulos, T. (2020b), "Blockchain technology for enhancing swift-trust, collaboration and resilience within a humanitarian supply chain setting." *International Journal of Production Research*, Vol. 58, pp. 3381–3398.

Dubey, R., Gunasekaran, A., Childe, S.J., Fosso Wamba, S., Roubaud, D. and Foropon, C. (2021b), "Empirical investigation of data analytics capability and organizational flexibility as complements to supply chain resilience." *International Journal of Production Research*, Vol. 59, pp. 110–128.

Erol, I., Oztel, A., Peker, I., Ar, I. M. and Searcy, C. (2024), "Critical success factors for building resilience in circular supply chains of electric vehicle batteries: Evidence from an emerging country." *Engineering Management Journal*, Vol. 38, pp. 1–19.

Ghomi, V., Gligor, D., Parast, M., Shokoohyar, S. and Esfahani, M.G. (2021), "Antecedents and consequences of customer flexibility: Establishing the link to firm competitive advantage." *Journal of Retailing and Consumer Services*, Vol. 62, p. 102609.

Gligor, D., Gligor, N., Holcomb, M. and Bozkurt, S. (2019), "Distinguishing between the concepts of supply chain agility and resilience." *The International Journal of Logistics Management*, Vol. 30, pp. 467–487.

Godwin-Geyi, D.A., Yusuf, Y., Menhat, M.S., Abubakar, T. and Ogbuke, N.J. (2020), "Agile capabilities as necessary conditions for maximising sustainable supply chain performance: An empirical investigation." *International Journal of Production Economics*, Vol. 222, p. 107501.

Gunasekaran, A., Yusuf, Y.Y., Adeleye, E.O. and Papadopoulos, T. (2018), "Agile manufacturing practices: The role of big data and business analytics with multiple case studies." *International Journal of Production Research*, Vol. 56, pp. 385–397.

Gunasekaran, A., Yusuf, Y.Y., Adeleye, E.O., Papadopoulos, T., Kovvuri, D. and Geyi, D.A.G. (2019), "Agile manufacturing: an evolutionary review of practices." *International Journal of Production Research*, Vol. 57, pp. 5154–5174.

Han, Y., Chong, W.K. and Li, D. (2020), "A systematic literature review of the capabilities and performance metrics of supply chain resilience." *International Journal of Production Research*, Vol. 58, pp. 4541–4566.

Ivanov, D. (2020), "Viable supply chain model: Integrating agility, resilience and sustainability perspectives-lessons from and thinking beyond the COVID-19 pandemic." *Annals of Operations Research*, Vol. 319, pp. 1–21.

Ivanov, D. (2021), "Lean resilience: AURA (active usage of resilience assets) framework for post-COVID-19 supply chain management." *The International Journal of Logistics Management*, Vol. 33, pp. 1196–1217.

Ivanov, D. and Dolgui, A. (2020), "Viability of intertwined supply networks: Extending the supply chain resilience angles towards survivability. A position paper motivated by COVID-19 outbreak." *International Journal of Production Research*, Vol. 58, pp. 2904–2915.

Jajja, M.S.S., Chatha, K.A. and Farooq, S. (2018), "Impact of supply chain risk on agility performance: Mediating role of supply chain integration." *International Journal of Production Economics*, Vol. 205, pp. 118–138.

Kamalahmadi, M. and Parast, M. (2016), "A review of the literature on the principles of enterprise and supply chain resilience: Major findings and directions for future research." *International Journal of Production Economics*, Vol. 171, pp. 116–133.

Kamalahmadi, M., Shekarian, M. and Parast, M. (2021), "The impact of flexibility and redundancy on improving supply chain resilience to disruptions." *International Journal of Production Research*, Vol. 60, pp. 1–29.

Kochan, C.G. and Nowicki, D.R. (2018), "Supply chain resilience: A systematic literature review and typological framework." *International Journal of Physical Distribution & Logistics Management*, Vol. 48, pp. 842–865.

Li, Y. and Zobel, C.W. (2020), "Exploring supply chain network resilience in the presence of the ripple effect." *International Journal of Production Economics*, Vol. 228, p. 107693.

Nascimento, D.L.M., Garcia-Buendia, N., Moyano-Fuentes, J. and Maqueira, J.M. (2024), "Unlocking the potential of industry 4.0 for supply chain flexibility and agility: A systematic literature review." *Engineering Management Journal*, Vol. 37, pp. 433–451.

Nath, V. and Agrawal, R. (2020), "Agility and lean practices as antecedents of supply chain social sustainability." *International Journal of Operations & Production Management*, Vol. 40, pp. 1589–1611.

Nemkova, E. (2017), "The impact of agility on the market performance of born-global firms: An exploratory study of the 'Tech City' innovation cluster." *Journal of Business Research*, Vol. 80, pp. 257–265.

Polyviou, M., Croxton, K.L. and Knemeyer, A.M. (2019), "Resilience of medium-sized firms to supply chain disruptions: The role of internal social capital." *International Journal of Operations & Production Management*, Vol. 40, pp. 68–91.

Queiroz, M.M., Ivanov, D., Dolgui, A. and Wamba, S.F. (2020), "Impacts of epidemic outbreaks on supply chains: Mapping a research agenda amid the COVID-19 pandemic through a structured literature review." *Annals of Operations Research*, Vol. 319, pp. 1–38.

Raji, I.O., Shevtshenko, E., Rossi, T. and Strozzi, F. (2021), "Industry 4.0 technologies as enablers of lean and agile supply chain strategies: An exploratory investigation." *The International Journal of Logistics Management*, Vol. 32, pp. 1150–1189.

Raut, R.D., Mangla, S.K., Narwane, V.S., Dora, M. and Liu, M. (2021), "Big data analytics as a mediator in lean, agile, resilient, and green (LARG) practices effects on sustainable supply chains." *Transportation Research Part E: Logistics and Transportation Review*, Vol. 145, p. 102170.

Sanchez, L.M. and Nagi, R. (2001), "A review of agile manufacturing systems." *International Journal of Production Research*, Vol. 39, pp. 3561–3600.

Santoro, G., Bertoldi, B., Giachino, C. and Candelo, E. (2018), "Exploring the relationship between entrepreneurial resilience and success: The moderating role of stakeholders' engagement." *Journal of Business Research*, Vol. 119, pp. 1–23.

Tukamuhabwa, B.R., Stevenson, M., Busby, J. and Zorzini, M. (2015), "Supply chain resilience: Definition, review and theoretical foundations for further study." *International Journal of Production Research*, Vol. 53, pp. 5592–5623.

Upadhyay, A., Mukhuty, S., Kumari, S., Garza-Reyes, J. A. and Shukla, V. (2022). "A review of lean and agile management in humanitarian supply chains: Analysing the pre-disaster and post-disaster phases and future directions." *Production Planning & Control*, Vol. 33, pp. 641–654.

Vastag, G., Kasarda, J.D. and Boone, T. (1994), "Logistical support for manufacturing agility in global markets." *International Journal of Operations & Production Management*, Vol. 14, pp. 73–85.

Wamba, S.F. and Akter, S. (2019), "Understanding supply chain analytics capabilities and agility for data-rich environments." *International Journal of Operations & Production Management*, Vol. 39, pp. 887–912.

2 Customer Relationship Management Excellence in Manufacturing Supply Chains

A Multilevel Analysis Using the ISCPM Framework

Ferdoush Saleheen and Md. Mamun Habib

2.1 INTRODUCTION

Owing to evolving customer demands, market fluctuations, and global supply chain disruptions, it has become increasingly demanding for manufacturers to keep pace in today's global competition and the era of digitalisation (Khourshed & Beshr, 2024; Wetsandornphong et al., 2025). Businesses are required to implement targeted customer strategies to promote flexibility, openness, and responsiveness to stay competitive. Therefore, customer relationship management (CRM) has become a critical instrument for enabling the internal supply chain to meet the expectations of the external market (Saleheen & Habib, 2022b). CRM is also regarded as a core and pivotal component of a broader strategic (ST) discipline that influences faster customer support, process coordination, and overall integration within the supply chain management (SCM) in an organisation (Chowdhury et al., 2025).

Additionally, CRM has become increasingly important in the current context of SCM research because it fosters resilience (RE), enhances cross-organisational collaboration (CL), and promotes transparency throughout the supply chain (Agyabeng-Mensah et al., 2025). There is not enough research done on CRM performance measurements through the analytic hierarchy process (AHP) and structural equation modelling (SEM). Therefore, there is a strong need for research on this topic for academics and industry practitioners in an emerging country like Bangladesh (Alkhawaldeh, 2025). This study utilised the Integrated Supply Chain Performance Measurement (ISCPM) model, which integrates the Supply Chain Operations Reference (SCOR) and Balanced Scorecard (BSC) frameworks (Saleheen et al., 2018a). This ISCPM model integrates CRM, internal supply chain management (ISCM), and supplier relationship management (SRM), and these components are examined through ST, planning (PL), and operational (OP) levels (Saleheen &

DOI: 10.1201/9781003770008-3

Habib, 2022a). The study highlights CRM measurement perspectives, with specific attention to how critical attributes such as RE, CL, and visibility (VS) support the enhancement of CRM. Recently, the importance of conducting empirical research in this area has increased considerably (Miraz et al., 2023; Saleheen & Habib, 2023b).

Despite the expanding recognition of CRM within SCM, prior studies largely conceptualise CRM as a supportive or technology-driven subsystem, unable to capture the distinctive role across ST, PL, and OP decision levels. This tendency has generated disjointed performance metrics that mask how CRM capabilities shape RE, CL, and VS in volatile supply chain environments. RE allows companies to tolerate disruptions and maintain service continuity, CL promotes shared goals and values and co-creation among supply chain partners, and VS gives real-time access to the decision-makers on accurate information to respond to market changes (Paul & Saha, 2025). There are few studies available for the researchers and practitioners to understand how these characteristics interrelate with one another at various decision levels of CRM and how they contribute to overall performance evaluation.

The AHP and SEM are both integrated into this study, where the AHP is applied to assess the relative significance of performance attributes, and the SEM is used to examine the interrelationships among attributes within CRM at ST, PL, and OP levels. Therefore, the purpose of this study is to fill a gap in the existing body of research. This study reinforces the debate in academia by empirically demonstrating the diverse capacity of CRM within the ISCPM framework, further augmenting the theoretical unification of CRM with supply chain performance measurement frameworks, and proposing a reproducible methodology for performance assessment in resource-constrained environments. These findings offer supply chain managers, individuals making decisions, and industrial planners recommendations so that they can use them to make their businesses more resilient and responsive to customers (Agyabeng-Mensah et al., 2025; Omoush, 2025).

Grounded in performance measurement theory and dynamic capability perspectives, this study positions CRM as an integrative capability rather than a downstream customer-facing function. The main objective of this research is to apply the ISCPM model to assess how effectively CRM performs at the ST, PL, and OP levels in manufacturing supply chain operations (Saleheen et al., 2018a).

The research specifically aims to:

- Employ the AHP to prioritise significant attributes of a CRM (Saleheen et al., 2018a).
- Implement SEM to validate the structural links between the CRM attributes (Khourshed & Beshr, 2024).
- Evaluate the importance of RE, CL, and VS in CRM performance in strengthening overall supply chain excellence (Agyabeng-Mensah et al., 2025; Paul & Saha, 2025).

And subsequently, the research questions are:

- How does CRM performance demonstrate ST-, PL-, and OP-level decisions (Saleheen & Habib, 2023b)?

Which CRM attributes have the most significance for achieving performance excellence within ISCPM (Alkhatib & Momani, 2023; Omoush, 2025)?

How do RE, CL, and VS perform together to make CRM more successful in the manufacturing sector in Bangladesh (Paul & Saha, 2025; Khedr, 2024)? The study recommends a hybrid strategy that applies AHP for attribute prioritisation through ranking and SEM for empirical verification. This approach ensures both conceptual accuracy and statistical reliability (RL), which are occasionally integrated into CRM research within the span of SCM (Saleheen et al., 2018a; Khourshed & Beshr, 2024). The study is done based on the Bangladesh manufacturing sector and is not applicable to the service industries such as retail, hospitals, or hospitality.

Section 2.2 presents a comprehensive review of the literature on CRM and its related outcome measures. Section 2.3 explains how the study was done, including the research method; how the data were collected; and how they were analysed. In Section 2.4, the model test results in AHP and SEM and their significance are explained. Section 2.5 discusses theoretical implications and practical applications, and finally, in Section 2.6, the conclusion summarises the research objectives and research questions, highlights the key findings, summarises how the findings impact the theory and practice, and then recommends prospective future study.

2.2 LITERATURE REVIEW

While these studies establish the relevance of CRM within supply chains, they largely stop short of explaining how CRM capabilities operate differently across organisational decision levels.

2.2.1 Supply Chain Macro Process

According to Chopra and Meindl, the Supply Chain Macro Process is made up of three parts: managing relationships with suppliers, managing internal operations, and managing relationships with customers (Chopra & Meindl, 2016). This provides any business with a better idea of how to acquire raw materials, process them, and send completed goods to clients. SRM is about how a company manages and cares for its suppliers so that they are productive and in line with the company's goals and key performance indicators (KPIs) (Saleheen et al., 2018a, 2023). Acquiring resources, coordinating logistics with suppliers, and collaborating to enhance quality are all integral parts of the overall process. ISCM examines every dimension of a firm's operations to ensure they can fulfil customer demand (Saleheen & Habib, 2023a).

The process involves managing production, inventory, scheduling, and ensuring OP continuity. Finally, CRM focuses on managing all customer-related activities, including order processing, demand forecasting, handling returns, and delivering high-quality services to achieve customer satisfaction (Chowdhury et al., 2025).

2.2.2 Evolution of Customer Relationship Management in Supply Chains

CRM has generally been presented within the domains of marketing and sales, emphasising the acquisition, retention, and development of customer relationships (Alkhawaldeh, 2025; Miraz et al., 2016). However, this old perception has been shifted over the last several years, and now the concept is viewed as a critical aspect of supply chain performance. This trend is because supply chains are focusing more on customers and using more digital technologies (Agyabeng-Mensah et al., 2025). Today, research shows that CRM is not only about sales but rather about predicting demand, delivering orders, providing customer service, and ensuring seamless service after the sales are all inclusive of the integrated CRM (Saleheen et al., 2014). Unfortunately, it does not reflect business models or all academic research on its application side, highlighting the urgency to design an application-based CRM model that is more focused on performance and compatible with how SCM overall operates (Khourshed & Beshr, 2024). This evolution reveals a persistent gap: CRM research has advanced conceptually but remains under-theorised as a performance-enabling capability within integrated supply chain systems.

2.2.3 CRM and Supply Chain Performance Measurement Frameworks

The CRM system is used by marketing and sales teams to recognise, maintain, and cultivate connections with customers (Chowdhury et al., 2025). The relevance of CRM has increased over the last several years. It is now believed that CRM is crucial to improving supply chains as they become more customer-focused and digital. In manufacturing, it is essential to accurately forecast demand, effectively manage orders, provide service, and retain ties with customers after sales have been completed (Saleheen et al., 2014). Traditional KPI-driven CRM models emphasise outcome indicators such as customer satisfaction and retention, yet they fail to capture the underlying OP and ST capabilities that enable sustained performance under disruption.

This marketing-centric orientation limits CRM's explanatory power in supply chain contexts characterised by OP volatility and inter-organisational dependencies. A considerable number of research and business CRM models feature actions that are solely focused on the customer, although this is not the case. There is a school of thinking that maintains that customer happiness and loyalty are not indicators of the true capabilities and processes of a company that contribute to the production of long-term value. The operations of the internal supply chain need to be gradually coupled with the CRM system, which is a tool for managing relationships with customers (Omoush, 2025).

2.2.4 Resilience in Customer-Centric Supply Chains

The ability of a company to anticipate, absorb, and recover from disruptions in the supply chain while simultaneously maintaining continuity of operations and levels of customer service is referred to as supply chain RE (Paul & Saha, 2025).

This capability enables the company to maintain both levels of customer service and continuity of operations. In the realm of CRM, RE exemplifies a business's ability to sustain its trust, service delivery, and customer engagement amid a crisis (Agyabeng-Mensah et al., 2025). Customer interaction strategies include risk monitoring, flexible resource allocation, and scenario PL, all of which are components of ST-level RE. Additionally, these components are all contained in the strategies (Khourshed & Beshr, 2024).

2.2.5 Collaboration as a Planning-Level CRM Enabler

CL may be described as the degree to which a company and its supply chain partners engage in cooperative PL, information exchange, and the alignment of their goals with one another, respectively (Wetsandornphong et al., 2025). CL in the CRM sector refers to the coordination of demand PL, the sharing of forecasts, and the joint design of services with customers and downstream partners (Omoush, 2025). Companies are able to incorporate consumer feedback directly into production and delivery plans via the use of collaborative CRM, which in turn fosters agility and co-created value (Agyabeng-Mensah et al., 2025).

2.2.6 Visibility and Operational-Level CRM Execution

According to Caridi et al. (2010), VS in supply chain operations refers to the capability of accessing, monitoring, and acting upon data that are being collected in real time throughout the whole value chain. VS that is generated by CRM guarantees that client preferences, orders, and service-level expectations are apparent to internal teams (Alkhatib & Momani, 2023). This enables a quicker response and an improvement in service quality. Technologies like enterprise resource planning (ERP), IoT, and customer analytics tools have made VS simpler, but a CRM strategy is needed to fully benefit from these breakthroughs (Khedr, 2024).

2.2.7 The ISCPM Model and CRM Attribute Integration

Three major theoretical contributions come from the ISCPM CRM dimension. It proves that CRM strategy and supply chain skills are linked. It quantifies the impact of CRM constructs (RE, CL, VS) using AMOS (Analysis of Moment Structure)-based SEM modelling. It provides empirical generalisability within the context of an emerging manufacturing economy (Saleheen & Habib, 2023b).

The ISCPM model formulated in this study is predicated on the theoretical frameworks of the BSC (Kaplan & Norton, 1996) and the SCOR model (SCC, 2012). When it comes to locations and industries, there is still a significant divide between the two. There is a significant quantity of research that originates from industrialised countries about CRM and supply chain effectiveness. Both the supply networks and CRM systems in these economies are technologically advanced and perform effectively together (Saleheen et al., 2018b).

Although BSC and SCOR frameworks offer structured performance perspectives, neither explicitly models CRM as a multilevel decision construct, creating a theoretical blind spot addressed by the ISCPM framework.

Despite the fact that emerging economies have their own challenges, such as a lack of infrastructure, a shortage of trained labour, and confusing laws, they are still not provided with enough representation. This is especially true for the manufacturing sector, which is a key part of industrial growth in many developing countries, such as Bangladesh (Saleheen & Habib, 2022a).

In conclusion, the literature identifies five significant research deficiencies: (I) The absence of CRM integration within multilevel supply chain performance frameworks; (II) the lack of real-world studies that connect CRM features to ST, PL, and OP decision levels; (III) the insufficient application of hybrid methodologies like AHP–SEM in CRM performance research; (IV) the inadequate examination of essential CRM attributes (RE, CL, VS) in a structured and comparative framework; and (V) the geographical bias favours developed economies, resulting in a lack of emphasis on CRM excellence within the manufacturing supply chains of emerging markets. This study addresses these gaps by using a hybrid AHP–SEM technique to evaluate CRM performance within the context of the ISCPM framework.

The research draws on data from 207 manufacturing enterprises in Bangladesh. This research is one of the first studies to characterise CRM qualities across several decision levels. It provides a framework that is both theoretically sound and practically usable, with the goal of attaining CRM excellence in supply chain settings that are still in their infancy (Saleheen & Habib, 2022b).

Synthesising the literature reveals five interrelated but unresolved gaps that collectively constrain CRM performance research. The present study addresses these gaps by empirically operationalising CRM attributes across decision levels within a unified ISCPM framework using a hybrid AHP–SEM methodology.

2.3 RESEARCH METHODOLOGY

A self-administered questionnaire was utilised in this exploratory study to gather both primary and secondary data. The study employed a quantitative research design, in which data were simultaneously collected and analysed concurrently to address the research objectives (Alkhawaldeh, 2025). The empirical setting of the study is the manufacturing sector in Bangladesh, covering 24 manufacturing sectors (Saleheen & Habib, 2022a).

The unit of analysis comprised manufacturing companies in Bangladesh, and the data were obtained from supply chain professionals via an email survey. Simple random sampling was implemented to achieve equal representation across the target population. The total population included 7,570 supply chain professionals. Based on a 95% confidence level, a ±7% margin of error, and the specified population size, the minimum required sample size was calculated to be 199 respondents (Saleheen & Habib, 2023a). Consistent with established methodological guidance, one SCM professional from each firm was designated to represent the organisation's supply chain function (Creswell & Creswell, 2018; Kothari, 2019; Saleheen et al., 2018a).

The questionnaire consisted of ten supply chain performance attributes measured through 36 performance measurement indicators. The 36 indicators functioned as independent variables, while the ten performance attributes functioned as dependent variables. A total of 1,832 supply chain professionals were approached via email, yielding 219 completed responses, corresponding to a response rate of approximately 12%. Following data validation, 12 responses were classified as inconsistent and were removed for further analysis. In addition, response bias and discrimination were assessed and found to be statistically insignificant (Creswell & Creswell, 2018; Kothari, 2019). Consequently, 207 valid responses were retained for subsequent analysis.

AHP was applied to prioritise supply chain and CRM performance drivers based on expert judgement, while SEM was employed to empirically validate the structural relationships among latent constructs. The integration of subjective prioritisation with objective statistical validation provides a robust methodological framework, ensuring both conceptual clarity and empirical robustness in evaluating supply chain and CRM performance (Creswell & Creswell, 2018; Kothari, 2019). The methodological design conforms to the positive research paradigm, emphasising objective measurement, statistical validation, and replicability. The integration of AHP and SEM facilitates the alignment of expert judgement with empirical validation, addressing long-standing critiques of purely subjective or solely statistical CRM evaluation approaches.

2.4 RESULTS AND DISCUSSIONS

2.4.1 Results of AHP Prioritising CRM Attributes

The ten CRM attributes were analysed at the ST, PL, and OP levels using the AHP technique. AHP has implemented expert pairwise comparisons that were compiled from 207 respondents in 24 manufacturing sectors in Bangladesh. The judgements were dependable, as evidenced by the fact that the consistency ratio for all matrices was less than 0.10 (Saaty, 1980; Creswell and Creswell, 2018; Kothari, 2019). Using the AHP method, the ten CRM attributes were looked at from the ST, PL, and OP points of view. AHP has used expert pairwise comparisons based on answers from 207 people in 24 manufacturing sectors in Bangladesh. The judgements were reliable, as shown by the fact that the consistency ratio for all matrices was less than 0.10.

In Table 2.1, the following abbreviations are followed: strategic (ST), planning (PL), operational (OP), resilience (RE), RL, work people health (WPH), sustainability (SS), CL, financial health (FH), velocity (VC), visibility (VS), continuous improvement (CI), and service excellence (SE) (Saleheen & Habib, 2022a).

The AHP results reveal that RE is the most dominant ST-level driver (RE = 0.537), followed by RL (RL = 0.243). At the PL level, CL emerged as the strongest attribute (CL = 0.352), surpassing VC (VC = 0.329). On the OP level, VC recorded the highest attribute (VC = 0.399), followed by CI (CI = 0.329) (Saleheen et al., 2018a).

These prioritisation patterns indicate that CRM effectiveness is contingent upon decision-level specificity rather than uniform attribute importance.

TABLE 2.1
Prioritising CRM Attributes

Level	Attribute	AHP Weight
ST	RE	0.537
ST	RL	0.243
ST	WPH	0.130
ST	SS	0.090
PL	CL	0.352
PL	FH	0.318
PL	VC	0.329
OP	VS	0.399
OP	CI	0.329
OP	SE	0.272

2.4.2 Overview of Structural Model Output

The CRM component of the ISCPM framework was tested using SEM in AMOS (Model III), evaluating the impact of ST (CRM_ST), PL (CRM_PL), and OP (CRM_OP) CRM levels on key performance outcomes. Both the model's validity and its fitness are shown to be strong by the outcomes (Saleheen & Habib, 2022b; Khourshed & Beshr, 2024).

The CRM attributes of the ISCPM model were validated through the SEM in AMOS. The purpose of this study was to investigate the role that ST (CRM_ST), PL (CRM_PL), and OP (CRM_OP) CRM elements played in crucial performance outcomes. According to the findings, the model is quite reliable.

2.4.3 Standardised Regression Weights

The standardised regression weights from CRM_ST reveal that RE in Table 2.2 is the most dominant ST-level driver ($\beta = 0.663$), followed by RL (RL = 0.600) and WPH (WPH = 0.599). When it comes to the PL level, CL emerged as the most powerful attribute ($\beta = 0.869$), surpassing both VC (VC = 0.615) and FH (FH = 0.303). Regarding the OP level, VS was found to have the greatest attribute ($\beta = 1.141$), followed by CI (CI = 0.522) and SE (SE = 0.126) (Saleheen & Habib, 2022a).

According to the standardised regression weights established between CRM_ST, it is evident that RE is the most significant ST-level driver, with a coefficient of 0.663. The next items on the agenda are WPH (WPH = 0.599) and RL (RL = 0.600). Following CL as the most powerful construction at the PL level were VC (VC = 0.615) and FH (FH = 0.303). CL had a beta value of 0.869, making it the most powerful construct. Within the realm of operations, VS had the highest significance ($\beta = 1.141$), followed by CI (CI = 0.522) and SE (SE = 0.126) (Paul & Saha, 2025).

TABLE 2.2
Standardised Regression Weights (CRM – ISCPM Model III)

CRM Construct	Latent Variable	Standardised Estimate (β)
CRM_ST → RE	0.663	$(p < 0.001)^{***}$
CRM_ST → RL	0.600	$(p < 0.001)^{***}$
CRM_ST → WPH	0.599	$(p < 0.001)^{***}$
CRM_ST → SS	0.128	$(p < 0.001)^{***}$
CRM_PL → CL	0.869	$(p < 0.001)^{***}$
CRM_PL → VC	0.615	$(p < 0.001)^{***}$
CRM_PL → FH	0.303	$(p < 0.001)^{***}$
CRM_OP → VS	1.141	$(p < 0.001)^{***}$
CRM_OP → CI	0.522	$(p < 0.001)^{***}$
CRM_OP → SE	0.126	$(p < 0.001)^{***}$

*** means p-value of less than 0.001.

2.4.4 Path Significance and Fit Indices

All regression paths were found to be statistically significant at $p < 0.001$, with critical ratios (CRs) above 3.0. Table 2.2 presents a summary of regression paths and standardised estimates. All regression paths were found to be statistically significant at $p < 0.001$, with CRs above 3.0. Table 2.2 gives a brief overview of the regression paths and standardised estimates.

Model Fit Summary: The structural model in Table 2.3 demonstrated a very good fit based on precise metrics such as Chi-square = 1.544 (acceptable ≤3), Comparative Fit Index = 0.926, Tucker–Lewis Index = 0.813, Incremental Fit Index = 0.774, and Root

TABLE 2.3
R^2 – Explained Variance of CRM Attributes

Construct	R^2 (%)
Resilience (RE)	44.0
Reliability (RL)	36.0
WPH	35.9
Collaboration (CL)	75.6
Financial health (FH)	9.2
Velocity (VC)	37.9
Continuous improvement (CI)	27.3
Visibility (VS)	30.1
Service excellence (SE)	21.6
Sustainability (SS)	1.6

Mean Square Error of Approximation = 0.058 (within threshold ≤ 0.08). These metrics indicate the model's robustness and statistical RL in describing CRM success across tiers.

Explained variance (R^2) of key constructs: The model's squared multiple correlations (R^2) indicate the proportion of variance explained by exogenous constructs.

R^2 results are summarised in Table 2.3. The results demonstrate that the approach is durable and statistically reliable for CRM performance clarification at several levels (Khourshed & Beshr, 2024).

CL (CL = 75.6%) and RE (RE = 44%) explained the most, demonstrating their importance for CRM excellence (Agyabeng-Mensah et al., 2025). The strong model fit confirms the structural validity of CRM as a multilevel construct within supply chain performance theory.

2.5 DISCUSSION AND THEORETICAL CONTRIBUTIONS

The findings extend CRM theory by demonstrating that CRM performance emerges from differentiated capability configurations across ST, PL, and OP levels (Saleheen & Habib, 2022a): RE, CL, and VS. Strategically, RE was the most important attribute, with a coefficient of 0.663% and 40.0%. This study supports the literature that emphasises ST agility in customer involvement in changing market circumstances by emphasising the necessity to adjust to shocks. CL was the most significant attribute at the PL level, surpassing FH and VC ($\beta = 0.869$, $R^2 = 75.6\%$) (Saleheen & Habib, 2023a).

There must be trust, commitment, and open and aligned communication among the internal team members, and they must cooperate to predict productive customer encounters (Saleheen & Habib, 2022b). The OP level showed the greatest impact of VS ($\beta = 1.141$, $R^2 = 30.1\%$), enhancing CRM effectiveness and responsiveness with real-time access to customer needs, inventory, and service data. Current digital SCM research favours ERPs and track-and-trace systems (Saleheen et al., 2018b).

From a dynamic capabilities perspective, RE reflects sensing and reconfiguring capacities, CL represents relational integration, and VS operationalises information-processing capability.

Sustainability (1.6%) and FH (9.2%) have relatively low priority, indicating that, despite their significance, they may not be the primary focus of customer engagement strategies in manufacturing SCM in Bangladesh. As a consequence, it adds to the body of knowledge on supply chain performance assessments by proposing an ISCPM framework rather than a KPI-driven approach. This is performed by empirically verifying the CRM component of the ISCPM model. By recasting CRM as a system that promotes RE, the emphasis changes from customer retention to ST and OP adaptability, which is critical in developing economies where upheavals are common (Saleheen & Habib, 2022a).

Based on the findings, practitioners should be aware that it is of the utmost importance to prioritise investments in real-time VS tools and CL platforms as part of CRM upgrade programmes. ST PL teams should make it a priority to have joint PL meetings with major customers and seminars on building RE in order to guarantee that their relationship management techniques will continue to be successful in the future (Saleheen et al., 2018b).

Supply chain managers have to institutionalise CRM metrics that include not just customer happiness but also VS and RE results. In order to improve VS, executives in charge of information technology and digital must combine CRM systems with ERP, warehouse management, and predictive analytics. The establishment of collaborative platforms that allow manufacturing businesses to share knowledge, cooperatively handle difficulties, and evaluate their RE should be a priority for policymakers and industry groups.

Beyond managerial relevance, these insights advance supply chain performance measurement by repositioning CRM as a structurally embedded capability rather than an outcome-oriented function.

2.6 CONCLUSION

This research examines a critical deficit in literature through investigating CRM performance in the manufacturing sector, grounded on the AHP ranking and SEM findings (Khourshed & Beshr, 2024; Saleheen & Habib, 2022b). The application of the ISCPM framework demonstrates that CRM performance is significantly affected by multitiered enablers: RE and RL at the ST level, CL and FH at the PL level, and VS and CI at the OP level (Omoush, 2025; Saleheen et al., 2018a).

This research contributes in three specific ways. Therefore, this provides theoretical validation for CRM as a quantitative and integrative function across supply chain tiers (Saleheen et al., 2018b). It offers practical insights for supply chain executives aiming to improve CRM maturity via investment in RE skills, development of collaborative networks, and enhancement of supply chain VS using technology and analytics (Agyabeng-Mensah et al., 2025; Khedr, 2024).

The assimilation of CRM into supply chain PL has transformed from a deliberate decision to an ST need in a more unpredictable and fiercely competitive marketplace (Wetsandornphong et al., 2025).

The study contributes theoretically by reconciling CRM, supply chain performance measurement, and multilevel decision theory within a single empirically validated framework.

Future studies could investigate the association between CRM skills and external ecosystem variables, such as consumer preferences and regulatory changes. The process is undergoing a digital transformation (Alkhawaldeh, 2025).

The future study could explore incorporating the ISCPM model into the manufacturing as well as the service industries, and the geographic locations could include Asia, GCC countries, and other continents, which would provide comparative perceptions while improving the model's applicability and acceptability to a wider community (Agyabeng-Mensah et al., 2025).

This study is limited by its focus on manufacturing firms in a single emerging economy and by reliance on cross-sectional survey data. Future research may extend longitudinal validation and cross-industry comparison.

This research develops a novel CRM performance evaluation paradigm. In turbulent environments with frequent disruptions and shifting customer demand, building RE capacity, networks of CL, and supply chain VS through technology and analytics is essential for CRM excellence (Omoush, 2025; Paul & Saha, 2025).

REFERENCES

Agyabeng-Mensah, Y., Oloruntoba, R., Earnest, J., & Mohammadi, H. (2025). Sustainable supply chain management and performance outcomes: Supply chain practice view and mediated moderation perspectives. *Business Strategy and the Environment*, 34(7), 8237–8258. https://doi.org/10.1002/bse.70023

Alkhatib, S. F., & Momani, R. A. (2023). Supply chain resilience and operational performance: The role of digital technologies in Jordanian manufacturing firms. *Administrative Sciences*, 13(2), 40. https://doi.org/10.3390/admsci13020040

Alkhawaldeh, A. E. (2025). Measuring supply chain performance. *Journal of Supply Chain Theory and Practice*, 10(4), 20–25. https://doi.org/10.55220/25766759.355

Caridi, M., Crippa, L., Perego, A., Sianesi, A., & Tumino, A. (2010). Do virtuality and complexity affect supply chain visibility? *International Journal of Production Economics*, 127(2), 372–383. https://doi.org/10.1016/j.ijpe.2009.08.016

Chopra, S., & Meindl, P. (2016). *Supply Chain Management: Strategy, Planning, and Operation*, 6th ed. Pearson.

Chowdhury, A. H. M. Y., Shahadat, M. M. H., Talukder, S. C., Csonka, A., & Fekete Farkas, M. (2025). Integrating CRM, lean practices, and use of IT to enhance operational performance: The mediating role of quality information sharing. *Logistics*, 9(3), 123. https://doi.org/10.3390/logistics9030123

Creswell, W. J., & Creswell, D. J. (2018). *Research Design: Qualitative, Quantitative, and Mixed Methods Approaches*, 5th ed. SAGE Publications Inc.

Kaplan, R. S., & Norton, D. P. (1996). *The Balanced Scorecard: Translating Strategy into Action*. Harvard Business School Press.

Khedr, A. S. S. (2024). Enhancing supply chain management with deep learning and machine learning techniques: A review. *Journal of Open Innovation: Technology, Market, and Complexity*, 10(3), 100379. https://doi.org/10.1016/j.joitmc.2024.100379

Khourshed, N., & Beshr, M. (2024). Evaluating supply chain performance measurement system (SCPMS) lifecycle during unexpected events. *International Journal of Logistics Systems and Management*, 12(1). https://doi.org/10.1080/21693277.2024.2345616

Kothari, C. R. G. (2019). *Research Methodology. Methods and Techniques* (p. 8, Ch. I.), 4th ed. New Age International Limited.

Miraz, M. H., Saleheen, F., Ashif, A. S. M., Hossain, M. A., Hasan, M. T., Hwang, H. J., & Kabir, A. (2023). The collaborative effort and efficiency of inquiry-based learning: Effect on teacher performance—The role of student performance. *International Journal of Online Pedagogy and Course Design*, 13(1), 1–23.

Miraz, M. H., Saleheen, F., & Rahman, M. (2016). *Supply chain management in service quality*. In *Proceedings of the International Conference on Industrial Engineering and Operations Management* (pp. 2097–2105). IEOM Society International. https://ieomsociety.org/ieom_2016/pdfs/621.pdf

Omoush, M. (2025 The impact of supply chain integration via mediator supply chain resilience on improvement in the performance of manufacturing sectors. *International Review of Management and Marketing*, 15(2), 157–170. https://doi.org/10.32479/irmm.18191

Paul, A., & Saha, S. C. (2025). A systematic literature review on flexible strategies and performance indicators for supply chain resilience. *Global Journal of Flexible Systems Management*, 26(1), 207–231. https://doi.org/10.1007/s40171-024-00415-x

Saaty, T. L. (1980). *The Analytic Hierarchy Process: Planning, Priority Setting, and Resource Allocation*. McGraw-Hill.

Saleheen, F., & Habib, M. M. (2022a). Constructing a performance measurement index for the manufacturing industry: An empirical study. *Open Journal of Social Sciences*, 10(3), 137–151. https://doi.org/10.4236/jss.2022.103010

Saleheen, F., & Habib, M. M. (2022b). Integrated supply chain performance measurement model for the manufacturing industry. *Journal of Service Science and Management*, 15(2), 55–70. https://doi.org/10.4236/jssm.2022.152005

Saleheen, F., & Habib, M. M. (2023a). Embedding attributes towards the supply chain performance measurement. *Cleaner Logistics and Supply Chain*, 7, 100090. https://doi.org/10.1016/j.clscn.2022.100090

Saleheen, F., & Habib, M. M. (2023b). *Supply chain performance measurement for the manufacturing industry: A study during the pandemic (COVID-19)*. In *AIP Conference Proceedings* (Vol. 2693, No. 1, p. 080013). https://doi.org/10.1063/5.0118758

Saleheen, F., Habib, M. M., & Hanafiz, Z. (2014). Challenges of warehouse operations: A case study in the retail supermarket. *International Journal of Supply Chain Management*, 3(4), 63–67.

Saleheen, F., Habib, M. M., & Hanafiz, Z. (2018a). An empirical study on supply chain management performance measurement through AHP. *International Journal of Supply Chain Management*, 7(3), 349–367.

Saleheen, F., Habib, M. M., & Hanafiz, Z. (2018b). Supply chain performance measurement model: A literature review. *International Journal of Supply Chain Management*, 7(3), 1–9.

Saleheen, F., Miraz, M. H., Ya'u, A., Mahmud, H., & Kabir, A. (2023). *Halal logistic use behaviour in Malaysia and the UAE: The role of religious faith, image, trust, facilities, performance, and consumer satisfaction*. In *Proceedings of the 2023 International Conference on Digital Applications, Transformation and Economy (ICDATE 2023)* (pp. 1–7). IEEE. https://doi.org/10.1109/ICDATE58146.2023.10248530

Supply Chain Council. (2012). *Supply Chain Operations Reference (SCOR) Model (Version 11.0)*. Supply Chain Council.

Wetsandornphong, S. T., Farr, R., Coleman, J., Salimi, Z., & Sweeney, E. (2025). Performance, integration, and dynamic capabilities in supply chains: An interpretive investigation of their relationships. *Supply Chain Management: An International Journal*, 30(4), 452–475.

3 The Role of Dynamic Capabilities in Building a Sustainable Supply Chain

Alica Grilec

3.1 INTRODUCTION

Supply chain management is fundamental to business competitiveness, efficiency, and customer satisfaction (Lummus & Vokurka, 1999). Businesses face a lot of disruptions, and the need to develop dynamic capabilities (DCs) became crucial for more resilient and sustainable supply chains (Aslam et al., 2020; Gligor et al., 2020; El Baz & Ruel, 2024), so integrating sustainability practices became an important driver of competitive advantage achievements (Razzak, 2023). To reach real sustainability goals, more than basic (one-time) operational improvement is required (Aslam et al., 2020; Rahman et al., 2024). In this context, DCs represent a company's ability to integrate, develop, and reconfigure internal and external resources to adjust to changing business conditions and environments (Teece et al., 1997; Linde et al., 2021). While the resource-based view (RBV) explains how companies achieve competitive advantage through valuable, rare, and distinctive resources (Barney, 1991; Barney, 2001), it does not fully discuss how companies should navigate changing environments. Even though recent studies show that DCs are important for companies' success and staying competitive (Cepeda & Vera, 2007; Samik & Al-Khashman, 2024), research focused on DCs is still fragmented (Weerawardena & Mavondo, 2011; Gligor et al., 2020). While some of the research focuses more on the resilience outcomes than on the viewing DCs in a more integrated way (Feizabadi et al., 2019; El Baz & Ruel, 2024), other sustainability-focused research studies often miss the view on the DCs as a whole (Alzate et al., 2024; Rahman et al., 2024).

Only a smaller portion of studies take into their research focus agility, adaptability, and alignment together, or their effort in sustainable transformation of supply chains (Feizabadi et al., 2019; Gligor et al., 2020). This study is significant as it helps to close the identified gap by exploring how the three DCs – agility, adaptability, and alignment – support the development of sustainable and resilient supply chains. The chapter will include the methodology overview, synthesis of the literature, thematic synthesis results, and conceptual framework, ending with discussion and conclusion.

 DOI: 10.1201/9781003770008-4

3.2 RESEARCH METHODOLOGY

This research uses a systematic literature review (SLR) approach to critically investigate the role of DCs, which contribute to building sustainable supply chains. A systematic approach was chosen because it allows clarity, validity, and auditability in synthesizing academic knowledge (Booth et al., 2012; Fink, 2014). This chapter follows the PRISMA (Preferred Reporting Items for Systematic Reviews and Meta-Analyses) 2020 framework (Page et al., 2021). The review covers studies published between January 1997 and October 2025. Searches were carried out in two major databases (Scopus and ScienceDirect) and supplemented by the search engine Google Scholar. The reviewed literature included peer-reviewed journal articles that examined at least one of the Triple-A capabilities (agility, adaptability, or alignment) in a supply chain context and linked these capabilities to sustainability and/or resilience outcomes. After the full-text screening, 62 studies were kept for the final synthesis, visually presented by the PRISMA 2020 flow diagram (Figure 3.1).

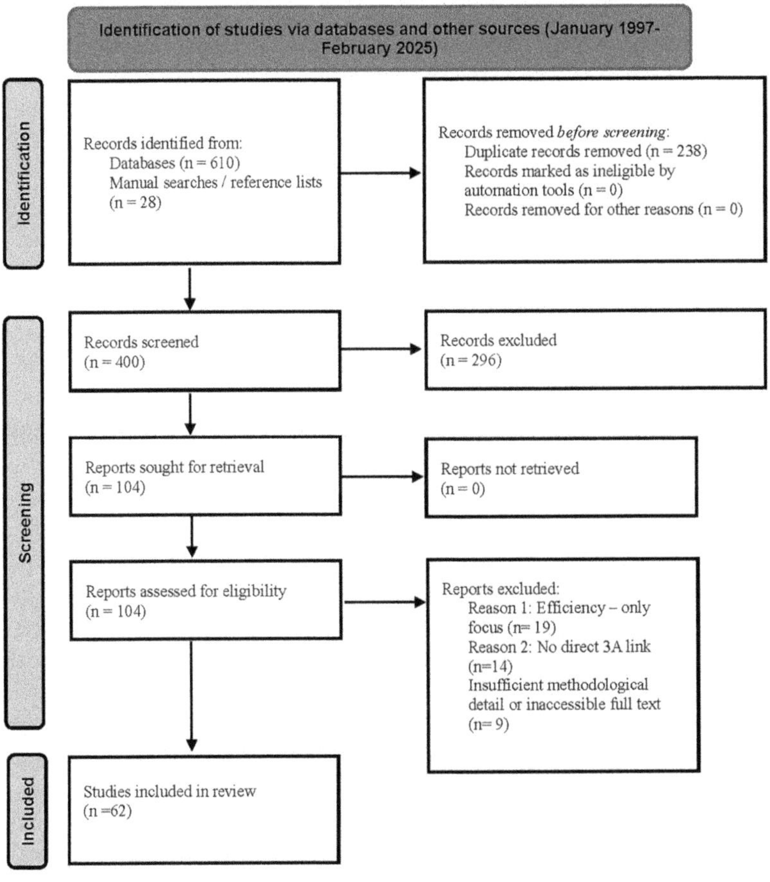

FIGURE 3.1 The PRISMA 2020 flow diagram of the study selection process.

(Source: Author, adapted from Page et al., 2021.)

The analysis involved two steps. First, the thematic synthesis was used (Thomas & Harden, 2008), and after that, the framework analysis (Ritchie & Spencer, 1994) was applied. As a result, three main themes were detected. All three themes are presented in Section 3.3, while the conceptual framework is presented in Section 3.4.

3.3 LITERATURE REVIEW

3.3.1 Theoretical Background

Starting with Barney (1991), companies' resources (assets, capabilities, processes, information, and knowledge) were viewed as resources that contribute to creating companies' competitive advantage when they are strategically combined and adapted to answer changing market conditions (Vozdvyzhenskyi & Titova, 2024). Recent empirical evidence further supports this view by showing that sustainable supply chain management practices are positively associated with competitive advantage and highlight the role of DCs in this relationship (Uddin et al., 2023).

When observed from the supply chain perspective, DCs refer to a company's ability to identify, use (apply), and adjust internal and external resources and information to improve overall supply chain operations (Wu et al., 2006). In addition, DCs are especially important in fast-changing, uncertain, and evolving supply chains where they help companies to adapt to the new situations by adjusting their operations (Wilden et al., 2013; Tuliakova et al., 2024). Defee and Fugate (2010) identified two DCs in the supply chain, knowledge assessing and co-evolving, that are important for supply chain development. Later, Beske (2012) expanded this work by adding sustainability-oriented DCs such as supply chain partner development, reflexive supply chain control, and supply chain re-conceptualization. Newest studies indicate that tools like artificial intelligence (AI) and predictive analytics can support resource configuration, while blockchain can strengthen interorganizational coordination and value creation, and as a result, shape how DCs operate in practice (Yuthas et al., 2021; Martínez-Pinzón & Duque, 2024; Roh & Xiao, 2024). Based on this understanding, the next section looks at the Triple-A framework – agility, adaptability, and alignment – as key DCs for successful and sustainable supply chains.

3.3.2 The Triple-A Framework in Sustainable Supply Chains

To build sustainable supply chains, efficiency alone is not enough, and three key DCs – agility, adaptability and alignment, often called Triple-A framework (introduced by Lee, 2004) – can also support sustainability by allowing companies to quickly react to disruptions, adjust fast to shifting demands, and coordinate their strategies across the supply chain, but these DCs were often treated in isolation (Feizabadi et al., 2019).

Research by Whitten et al. (2012) was the first research that tested Lee's (2004) theoretical background, bringing together agility, adaptability, and alignment and showing their impact on the supply chain performance. Li et al. (2015) and Gala-Velásquez and Arredondo-Salas (2019) linked agility, adaptability, and alignment with competitive performance. Aslam et al. (2020) advanced the Triple-A framework even further by conceptualizing adaptability and alignment and supply chain ambidexterity and linking them to resilience and adaptability. More recent studies further highlight the importance of Triple-A capabilities in helping supply chains to predict, respond, and recover from market disruptions, confirming their central role in building resilience (Alzate Rendón et al., 2025; Gomes & Silva, 2025). In addition, research shows that development and effectiveness of these DCs are supported by enabling factors such as data analytics, digital technologies, HR practices, leadership, and strategic alignment (Daneshvar Kakhki et al., 2023; Alfalla-Luque et al., 2025; Bag et al., 2025; Kiss, 2025)

3.3.2.1 Agility – Rapid Response to Market Changes and Disruptions

Agility is considered to be the ability of the supply chain and all its members to promptly coordinate all the actions with the shifting market needs, and react fast to the challenges, in addition to reacting to the customer needs, despite the unstable conditions on the market (Ismail & Sharifi, 2006; Lin et al., 2006; Motadel et al., 2011). Today, agile supply chains must also adapt quickly to new environmental goals, reduce carbon footprints, and adapt to regulatory and stakeholder pressures for greener supply chain operations (Shafaghizadeh & Sajadi, 2023). Thus, the idea of green supply chain agility was born, in which responsiveness is combined with environmental considerations such as sustainable materials, waste reduction, and responsible resource use (Khader & Nissar, 2025), enabling operations that are efficient and at the same time environmentally conscious (Saeed & Shoukat, 2023).

As Tabesh et al. (2024) argue, close supplier relationships help improve quality and support agility in sustainable initiatives. Market sensitivity, speed, and information sharing became key enablers (Agarwal et al., 2007). Digital tools such as Internet of Technology (IoT) further improve supply chain agility by improving the sharing of information faster, speeding up the decision process, and promoting more responsive operations (Schianchia, 2023).

From a critical perspective, agility is often recognized and seen as important DC, but its connection with sustainability is unclear as it is rarely observed from an environmental perspective. In previous research, agility is linked with being fast and responsive, and it can lead to increased resource use or even emissions if the sustainability aspect is not included. These gaps suggest two main issues in the existing literature: the impact of sustainability on agility often depends on digital technologies and collaborative practices, and agility is usually observed in isolation, without adaptability and alignment (Feizabadi et al., 2019).

3.3.2.2 Adaptability – Evolving With the Environment

Adaptability is a DC that gives supply chains the power to reshape themselves in response to long-term changes in their environment. It's about being proactive: redesigning systems and structures (networks) to stay ahead of the competition (Ketchen et al., 2008; Whitten et al., 2012). Research on agri-food supply chains showed how adaptability helps companies to manage risks caused by climate change and also economic instability, thereby supporting long-term resilience and sustainability (Alzate et al., 2024). In an unpredictable business environment, where demand shifts quickly and only uncertainty is constant, adaptability has become critical for supply chains to sustain long-term performance (Chan & Chan, 2010; Phadnis, 2023). To understand how adaptability works in practice, it is important to differentiate it from flexibility. Flexibility generally describes the internal capabilities of a supply chain, such as the ability to modify procurement, production, or distribution activities in response to particular operational needs, while adaptability includes long-term and strategic orientation, where the supply chain evolves in response to the changes within the market business environment (Ketchen et al., 2008; Whitten et al., 2012). As adaptability is closely linked to organizational culture, an adaptive green culture drives internal transformation of the companies, helping them strengthen sustainability practices and improve environmental performance (Mousa et al., 2025).

Today, adaptability involves rethinking and adopting technologies. To unlock this potential, companies need to integrate both technological tools and collaborative practices across their supply chain networks (Gligor et al., 2020). As Singh and Mathiyazhagan (2024) point out, adaptability supports smoother operational performance, which in turn, strengthens a company's ability to respond to customers' requirements in an unstable business environment. Modern supply chains rely on technological factors such as real-time data analytics, AI, and cloud-based platforms to navigate disruption, anticipate change, and coordinate strategies to keep responsiveness (Ivanov et al., 2010; Schianchia, 2023; Bag et al., 2025). In addition, keeping the supply chain simple has become one of the most effective ways to remain adaptable, as it makes supply chains more responsive and efficient (Hoole, 2005).

Strategic adaptability is crucial when facing disruptions or adapting to sustainability goals, as it enables companies to accept greener materials, cut down on waste, and launch eco-friendly products more effectively. As Phadnis (2023) notes, even well-intentioned sustainability plans can struggle when they do not have sufficient adaptability in unpredictable environments. From a critical perspective, even adaptability is strategically important, but it remains the least researched of the Triple-A capabilities. Existing studies are often context-specific; there is limited longitudinal evidence, and adaptability is sometimes mixed with flexibility, which blurs conceptual boundaries. While adaptability is associated with positive sustainability outcomes, few studies quantify these effects or focus on trade-offs, for example, as cost implications.

3.3.2.3 Alignment – Strategic Cohesion Across the Supply Chain

Alignment refers to the strategic coherence among all members of the supply chain through shared planning, clear roles between supply chain partners, and fair distribution of risk and reward across partners supported by harmonizing both external partnership that helps establish trust and collaboration throughout the supply chain (Lee, 2004; Towers & Burnes, 2008).

Previous research shows that collaboration is important in building sustainable and resilient supply chains, as buyer and supplier partnerships supported by trust improve traceability, environmental performance, resource and risk sharing, real-time information exchange, collective decision making, and coordinated resource utilization across supply chains, improving performance, speeding up delivery, and increasing customer satisfaction (Ragatz et al., 1997; Power et al., 2001; Martinez-Olvera, 2008; Tang & Tomlin, 2008; Duong et al., 2023; Boscari et al., 2024; dos

Santos et al., 2024; Nagy & Szentesi, 2024). Tools like CPFR further help alignment in planning and forecasting while contributing to sustainability goals, like environmental and social performance (Almario, 2024). When all partners in the supply chain have the same sustainability goals, they can make greener products and use circular practices more quickly. Alignment ensures that partners move in the same direction toward long-term sustainability, which also strengthens resilience and long-term value creation.

From a critical perspective, although alignment is widely recognized as a foundation for effective supply chain collaboration, its integration with sustainability goals is uneven across the literature. Studies tend to center on dyadic relationships (e.g. buyer supplier) and not on network level alignment. Little empirical research has been done on how trust and power relationships can be used to maintain alignment over time. The contribution of alignment to sustainability often occurs through digital integration and data sharing, and it brings the most benefits when developed alongside agility and adaptability, as complementary capabilities.

3.3.3 Thematic Synthesis Results

Following the three-stage approach of Thomas and Harden (2008), the 62 studies identified in the review were analyzed and synthesized. First, relevant findings, conclusions, and conceptual contributions were coded to capture explicit and implicit links between agility, adaptability, alignment, and sustainability in supply chain contexts. These codes were then grouped into clusters representing common patterns, mechanisms, or outcomes. Finally, analytical themes were developed to explain how the three capabilities interact, the conditions under which they produce sustainability benefits, and the limitations or contradictions present in the evidence base. The synthesis produced three overarching and interrelated themes.

3.3.3.1 Theme 1: Dynamic Capabilities as Networked, Multi-Level Phenomena

In literature, agility, adaptability, and alignment are always represented as not independent organizational qualities, but as network-based capabilities that are functioning at various levels of the supply chain. At the firm level, these capabilities are reflected in internal flexibility, faster decision, and the ability to adjust operations when conditions change. On the dyadic level, they manifest themselves by strong collaboration between buyers and suppliers, which is supported by common planning, sharing information, and cooperation in solving problems. On a broader network level, DCs are reinforced by broader collaborative structures, i.e., shared data platform and logistics partnering, which allow firms to pursue congruent sustainability goals.

3.3.3.2 Theme 2: The Enabling Role of Digital and Collaborative Practices

Second theme highlights the central role held by technological tools and partner relationships in facilitating Triple-A capabilities in producing sustainability outcomes. In the studies reviewed, digital systems (IoT, blockchain, predictive analytics, and cloud-based platforms) are determined as one of the key enablers that can help companies to track risks in real time, consider alternative options in a faster manner, and organize more effectively sustainability-related activities across the supply chain partners. In addition to these technologies, collaborative mechanisms, such as Collaborative Planning, Forecasting, and Replenishment (CPFR); joint innovation efforts; and shared sustainability scorecards facilitate alignment and increase the agility and adaptability impacts together. The synthesis suggests that sustainability is likely to be watered down or transient in the absence of such digital and collaborative infrastructures.

3.3.3.3 Theme 3: The Integrated Value of Agility, Adaptability, and Alignment

Even though each of the capabilities has its own impact on the supply chain resilience and sustainability, the review reveals that their combined and synchronized implementation produces the strongest and most sustainable results. Agility mainly facilitates quick/fast responsiveness in answer to the sustainability-related needs, including the swift transition to more environmentally friendly suppliers after regulation changes. Adaptability allows longer-term transformation to take place by facilitating structural and cultural changes in favor of more sustainable operating models. The complementary role of alignment is to align all the members of the supply chain toward the same strategic direction and implement sustainability into the common goals and daily practices.

The insights from synthesis suggest that the development of such capabilities alone leads to inconsistent sustainability gains. Agile responses can, as an example, be effective in reducing lead times but unintentionally higher carbon emissions in case they are accomplished by speedy freight. In the cases of the highest performance, indicated in the literature, the three capabilities are mutually supporting components: agility is used to comply immediately, adaptability is used to redesign a supply chain to become sustainable, and alignment is used to consolidate these alterations by collaborative structures of governance. The three themes together describe the way Triple-A capabilities are interdependent levers used to achieve sustainability of supply chain performance and indicate areas of empirical and conceptual gaps.

3.4 CONCEPTUAL FRAMEWORK

Based on the results of the thematic synthesis (Section 3.3), the conceptual framework (Figure 3.2) conceptualizes agility, adaptability, and alignment as mutually dependent DCs that, together, facilitate sustainable performance of supply chains. Resilient and sustainable supply chain performance thus becomes the constructive result of Triple-A, which is enabled through digital and collaborative infrastructures.

3.5 DISCUSSION

3.5.1 Interpretation of Findings

The findings of this systematic review confirm the fact that agility, adaptability, and alignment should be viewed as an interdependent set of DCs, as they generate more

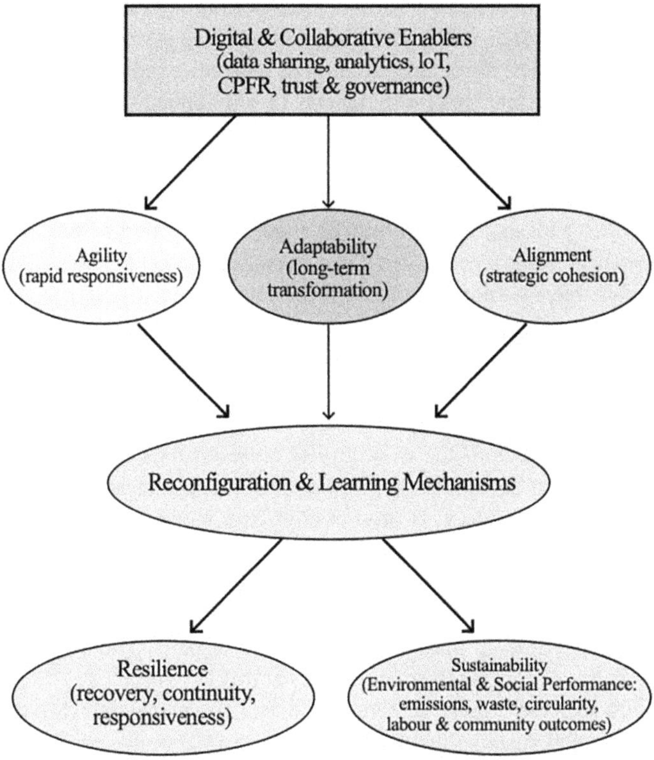

FIGURE 3.2 Conceptual framework.

(Source: Author.)

resilience and sustainability results when deployed jointly. This makes the Triple-A framework (Lee, 2004) even more relevant, while extending its explanatory potential to sustainable supply chain contexts.

The synthesis puts emphasis on the fact that each capability has a unique but complementary role. Agility primarily facilitates short-term responsiveness to disruptions, enabling supply chains to ensure operational continuity and respond quickly to shifting market, regulatory, and stakeholder pressures (Ismail & Sharifi, 2006; Lin et al., 2006). Nevertheless, the results also indicate that the benefits of agility can be achieved only temporarily unless it is accompanied by a wider coordination of the strategy.

Adaptability arises as the core transformational capability within the Triple-A model. In line with the DC theory (Teece et al., 1997), adaptability enables long-term structural, cultural, and technological reconfiguration of the supply chain networks. It enables companies to reform sourcing strategies, production systems, and partnership structures in response to continuing environmental uncertainty and sustainability pressures (Ketchen et al., 2008; Whitten et al., 2012). Alignment gives the strategic unity that is required to make agility and adaptability aimed at achieving common goals, and it enables the incorporation of sustainability objectives into joint planning, governance systems, and performance measurement systems (Duong et al., 2023; Boscari et al., 2024). The absence of alignment will cause rapid or adaptive responses to be fragmented, thus causing inefficiencies or unintended trade-offs across supply chain tiers. An enabling role in this process is played by digital technologies and collaborative mechanisms of governance. IoT, analytics, and blockchain assist in increasing sensing, coordination, and decision-making, whereas collaborative practices, like CPFR and shared performance metrics, help build trust and alignment within the partners (Gligor et al., 2020; Daneshvar Kakhki et al., 2023; Dickens et al., 2024). These enablers do not substitute the DCs but increase their effects by supporting the organizational learning and reconfiguration of resources.

3.5.2 Theoretical and Managerial Contributions

This research contributes to the existing literature on DCs and sustainable supply chain management in several ways. To begin with, it enlarges the RBV by demonstrating that valuable resources are not enough to achieve sustainability-oriented competitiveness. Second, the paper advances the Triple-A model by theorizing the relationships between agility, adaptability, and alignment to be a system of mutually supporting capabilities as opposed to single-mechanism views. Third, the research explains how the Triple-A framework should be organized internally by locating adaptability as the major transformational capability.

The results provide viable recommendations to managers who may wish to develop resilient and sustainable supply chains, as collaboration based on trust, joint planning, and common performance measures is critical in maintaining the alignment and incorporating sustainability objectives into the decision-making process. Overall, the integrated Triple-A framework will equip the managers with a systematic roadmap on how to balance responsiveness, resilience, and sustainability.

3.5.3 Research Gaps and Future Directions

Although growing interest is now paid to Triple-A capabilities, the interdependence between them is still rarely theorized or empirically tested, especially in the sustainability-based supply chain context, so integrated models that address the joint impacts of agility, adaptability, and alignment on sustainability and resilience outcomes should be developed. Second, sustainability is often measured using narrow efficiency-oriented indicators. Further studies should employ broader measurement frameworks that incorporate environmental, social, and governance dimensions to reflect comprehensive sustainable supply chain performance. Third, cross-sectional and single-industry research prevails, which restricts the understanding of the evolution of Triple-A capabilities over time, which requires longitudinal and mixed-method research designs. Lastly, the mediating roles of digital technologies and collaborative governance are unexplored, even though they are identified as enablers. Future research should investigate the interaction of digitalization and collaboration with Triple-A capabilities across multitier supply chain networks and evaluate possible trade-offs between responsiveness and strategic alignment.

3.6 CONCLUSION

The chapter explored the relationship between agility, adaptability, and alignment, and their interaction to support resilience and sustainability in global supply chains. Based on an SLR that was conducted using the PRISMA framework, the study reveals that Triple-A capabilities work best as a reinforcing system. The findings indicate the leading transformational role of adaptability, supported by agility's short-term responsiveness and alignment's strategic coherence, with digital technologies and collaborative governance acting as enabling factors.

The proposed conceptual framework integrates insights from RBV and DC theory and offers theoretical clarity and practical advice on how to create supply chains that are resilient to disruption and, at the same time, be responsive to the changing environmental and social expectations.

Despite these contributions, there are several limitations that can be pointed out. Only English-language, peer-reviewed papers are reviewed, which can develop publication bias and language bias as well. Moreover, cross-sectional and case-based studies are predominant, which limit the possibility of causal inference and restrict the understanding of how Triple-A capabilities develop and interact over time.

The proposed framework should be empirically tested in future studies using longitudinal and mixed-method research designs in order to capture the organizational learning, resource reconfiguration, and capability evolution processes. Scholars are encouraged to adapt broader sustainability indicators and metrics, including social, governance, and circular economy dimensions, as well as to examine Triple-A capability interrelations across multitier and global supply chain networks. Addressing these issues will further refine theory and support the development of competitive, resilient, and genuinely sustainable supply chains.

REFERENCES

Agarwal, A., Shankar, R., & Tiwari, M.K. (2007). Modeling agility of supply chain. *Industrial Marketing Management*, 36, 443–457. https://doi.org/10.1016/j.indmarman.2005.12.004

Alfalla-Luque, R., Marín-García, J. A., Machuca, J. A. D., & Díaz-Curbelo, A. (2025). Exploring the links between human resources and triple-A supply chain capabilities. *International Journal of Logistics Management*, 36(3), 968–986. https://doi.org/10.1108/IJLM-01-2024-0038

Almario, R. (2024). Risk management and collaboration on firms' sustainability for supply chain resilience. *International Journal of Research Studies in Management*, 12(12), 1–14. https://doi.org/10.5861/ijrsm.2024.1262

Alzate, I. C., López, Y. M., Manotas, E. C., & Boada, A. (2024). Flexibility and adaptability: Dynamic capabilities for building supply chain resilience. *Journal of Infrastructure, Policy and Development*, 8(16), 1–28. https://doi.org/10.24294/jipd9749

Alzate Rendón, I. C., Álvarez-Gallo, S. M., & Boada, A. J. (2025). Supply chain resilience through collaborative networks and dynamic capabilities: Evidence from an agri-food productive chain in Colombia. *Discover Sustainability*, 6(1). https://doi.org/10.1007/s43621-025-02094-9

Aslam, H., Khan, A. Q., Rashid, K., & Rehman, S. U. (2020). Achieving supply chain resilience: The role of supply chain ambidexterity and supply chain agility. *Journal of Manufacturing Technology Management*, 31(6), 1185–1204, https://doi.org/10.1108/JMTM-07-2019-0263

Bag, S., Rahman, M. S., Sharma, J. R., Chiarini, A., Shrivastav, S. K., & Gupta, S. (2025). Building digital technology and innovative lean management capabilities for enhancing operational performance: An empirical study. *Production Planning and Control*, 36(8), 1009–1028. https://doi.org/10.1080/09537287.2024.2327344

Barney, J. B. (1991). Firm resources and sustained competitive advantage. *Journal of Management*, 17, 99–120. https://doi.org/10.1177/014920639101700108

Barney, J. B. (2001). Resource-based theories of competitive advantage: A ten-year retrospective on the resource-based view. *Journal of Management*, 27, 643–650. https://doi.org/10.1177/014920630102700602

Beske, P. (2012). Dynamic capabilities and sustainable supply chain management. *International Journal of Physical Distribution & Logistics Management*, 42(4), 372–387. https://doi.org/10.1108/09600031211231344

Booth, A., Papaioannou, D., & Sutton, A. (2012) *Systematic Approaches to a Successful Literature Review*. Sage.

Boscari, S., van Donk, D.P., Pullman, M. and Xiao, C. (2024). Different paths to improving together: a taxonomy of buyer-supplier collaborations for sustainability in food supply networks. *Supply Chain Management*, 29(6), 996–1018. https://doi.org/10.1108/SCM-02-2024-0119

Cepeda, G., & Vera, D. (2007). Dynamic capabilities and operational capabilities: A knowledge management perspective. *Journal of Business Research*, 60, 426–437. https://doi.org/10.1016/j.jbusres.2007.01.013

Chan, H.K., & Chan, F.T.S. (2010). Comparative study of adaptability and flexibility in distributed manufacturing supply chains. *Decision Support Systems*, 28, 331–341. https://doi.org/10.1016/j.dss.2009.09.001

Daneshvar Kakhki, M., Rea, A. I., & Deiranlou, M. (2023). Data analytics dynamic capabilities for Triple-A supply chains. *Industrial Management and Data Systems*, 123(2), 534–555. https://doi.org/10.1108/IMDS-03-2022-0167

Defee, C.C., & Fugate, B.S. (2010). Changing perspective of capabilities in the dynamic supply chain era. *The International Journal of Logistics Management*, 21(2), 180–206. https://doi.org/10.1108/09574091011071915

Dickens, J. M., Uvet, H., Anderson, J. R., Ponomarov, S. Y., & Nowicki, D. (2024). An empirical examination on firm innovativeness and supply chain resilience: The role of information sharing from a dynamic capabilities view, *Knowledge and Process Management*, 31(4), 295–308 https://doi.org/10.1002/kpm.1785

Duong, L. N. K., Kumar, V., & He, Q. (2023). Collaboration for the sustainable food supply chain: A bibliometric analysis. *Business Strategy and the Environment*, 34(3), 1287–1301. https://doi.org/10.1002/bse.4051

El Baz, J., & Ruel, S. (2024). Achieving social performance through digitalization and supply chain resilience in the COVID-19 disruption era: An empirical examination based on a stakeholder dynamic capabilities view. *Technological Forecasting & Social Change*, 201, 123–209. https://doi.org/10.1016/j.techfore.2024.123209

Feizabadi, J., Maloni, M., & Gligor, D. (2019). Benchmarking the triple-A supply chain: Orchestrating agility, adaptability, and alignment. *Benchmarking: An International Journal*. https://doi.org/10.1108/BIJ-03-2018-0059

Fink, A. (2014) *Conducting Research Literature Reviews: From the Internet to Paper* (4th ed.). Sage.

Gala-Velásquez, B. D. La, & Arredondo-Salas, A. Y. (2019). The adaptability and alignment as predictive variables of the agility in the MYPES of the textile sector of the Arequipa region, Peru: An analysis with triple "A" approach; La adaptabilidad y alineamiento como variables predictores de la agilidad en las MYPES del sector textil de la región Arequipa, Perú: Un análisis con enfoque triple "A." *Espacios*, 40(28), 1–20.

Gligor, D., Feizabadi, J., Russo, I., Maloni, M. J., & Goldsby, T. J. (2020). The triple-a supply chain and strategic resources: Developing competitive advantage. *International Journal of Physical Distribution & Logistics Management*, 50(2), 159–190, https://doi.org/10.1108/IJPDLM-08-2019-0258

Gomes, P. J., & Silva, G. M. (2025). Navigating the interplay between digitalization and triple-A capabilities for enhanced supply chain resilience. *Journal of Business Logistics*, 46(3). https://doi.org/10.1111/jbl.70018

Hoole, R. (2005). Five ways to simplify your supply chain. *Supply Chain Management: An International Journal*, 10(1), 2–6. https://doi.org/10.1108/13598540510578306

Ismail, H. S., & Sharifi H. (2006). A balanced approach to building agile supply chains. *International Journal of Physical Distribution and Logistics Management*, 36(6), 431–444. https://doi.org/10.1108/09600030610677384

Ivanov, D., Sokolov, B., & Kaeschel, J. (2010). A multi-structural framework for adaptive supply chain planning and operations control with structure dynamics considerations. *European Journal of Operational Research*, 200, 409–420. https://doi.org/10.1016/j.ejor.2009.01.002

Ketchen, D. J., Rebarick, W., Hult, G. T. M., Meyer, D. (2008). Best value supply chains: A key competitive weapon for the 21st century. *Business Horizons*, 51, 235–243. https://doi.org/10.1016/j.bushor.2008.01.012

Khader, A. R., & Nissar, P. (2025). Impact of green structural capital on green supply chain agility. In M. Majeed, K. Agarwal, & A. Tijani (Eds.), *Green Supply Chain Management* (1st ed., p. 9). Apple Academic Press.

Kiss, E. M. (2025). Unpacking the micro-foundations of dynamic capabilities in retail banking: The roles of strategic alignment, leadership, and organisational structure. *International Journal of Organizational Analysis*. https://doi.org/10.1108/IJOA-10-2024-4920

Lee, H. L. (2004). The triple-A supply chain. *Harvard Business Review*, 82(10), 102–112. https://hbr.org/2004/10/the-triple-a-supply-chain

Li, X., Wu, Q., & Holsapple, C. W. (2015). Best-value supply chains and firms' competitive performance: Empirical studies of their linkage. *International Journal of Operations and Production Management*, 35(12), 1688–1709. https://doi.org/10.1108/IJOPM-01-2014-0014

Lin, C.T., Chiu, H., & Chu, P.-Y. (2006). Agility index in the supply chain. *International Journal of Production Economics*, 100, 285–299. https://doi.org/10.1016/j.ijpe.2004.11.013

Linde, L., Sjödin, D., Parida, V., & Wincent, J. (2021). Dynamic capabilities for ecosystem orchestration: A capability-based framework for smart city innovation initiatives. *Technological Forecasting & Social Change*, 166, 120614. https://doi.org/10.1016/j.techfore.2021.120614

Lummus, R.R., & Vokurka R.J. (1999). Defining supply chain management: A historical perspective and practical guidelines. *Industrial Management and Data System*, 99(2), 11–17. https://doi.org/10.1108/02635579910243851

Martinez-Olvera, C. (2008). Methodology for realignment of supply-chain structural elements. *Internal Journal of Production Economics*, 114, 714–722. https://doi.org/10.1016/j.ijpe.2008.03.008

Martínez-Pinzón, C. A., & Duque, P. (2024). Scientometric insights into dynamic capabilities: Measuring academic impact and trends. *CLIO América* 18(35), 106–125. https://doi.org/10.21676/23897848.5816

Motadel, M., Toloie-Eshlaghy, A., & Halvachi-Zadeh, D. (2011). Assessment of supply chain agility in the Automotive Industry of Teheran. *European Journal of Scientific Research*, 61(2), 210–229. https://doi.org/10.1016/J.IJPE.2008.03.008

Mousa, H., Khalifa, W., & Alzubi, A. (2025). Green supply chain practices and environmental performance: A moderated role of adaptive green culture and mediated role of competitive pressure. *Sustainability*, 17(12), 1–23. https://doi.org/10.3390/su17010012

Nagy, G., & Szentesi, S. (2024). Collaborative logistics: An innovative strategy to address future logistics challenges. *Advanced Logistic Systems-Theory and Practice*, 18(3), 83–95. https://doi.org/10.32971/als.2024.031

Page, M. J., McKenzie, J. E., Bossuyt, P. M., Boutron, I., Hoffmann, T. C., Mulrow, C. D., Shamseer, L., Tetzlaff, J. M., Akl, E. A., Brennan, S. E., Chou, R., Glanville, J., Grimshaw, J.M., Hróbjartsson, A., Lalu, M. M., Li, T., Loder, E. W., Mayo-Wilson, E., McDonald, S., McGuinness, L. A., Stewart, L. A., Thomas, J., Tricco, A. C., Welch, V. A., Whiting, P., Moher, D. (2021). The PRISMA 2020 statement: An updated guideline for reporting systematic reviews. *BMJ*, 372, n71. https://doi.org/10.1136/bmj.n71

Phadnis, S. (2023). A review of research on supply chain adaptability: Opening the black box. *Journal of Business Logistics*, 44(4), 1–30. https://doi.org/10.1111/jbl.12370

Power, D. J., Sohal, A. S., & Rahman, S.-U. (2001). Critical success factors in agile supply chain management. *International Journal of Physical Distribution and Logistic Management*, 31(4), 247–265. https://doi.org/10.1108/09600030110394923

Ragatz, G. L., Handfield, R. B., & Scannell, T. V. (1997). Success factors for integrating suppliers into new product development. *Journal of Product Innovation Management*, 14(3), 190–202. https://doi.org/10.1016/S0737-6782(97)00007-6

Rahman, M.A., Saha, P., Belal, H.M., Hasan Ratul, S., & Graham, G. (2024), Big data analytics capability and supply chain sustainability: Analyzing the moderating role of green supply chain management practices. *Benchmarking: An International Journal*. https://doi.org/10.1108/BIJ-10-2024-0852

Razzak, M. R. (2023). Mediating effect of productivity between sustainable supply chain management practices and competitive advantage: Evidence from apparel manufacturing in Bangladesh. *Management of Environmental Quality*, 34(2), 428–445. https://doi.org/10.1108/MEQ-01-2022-0022

Ritchie, J., & Spencer, L. (1994). Qualitative data analysis for applied policy research. In A. M. Huberman & M. B. Miles (Eds.), *The Qualitative Researcher's Companion* (pp. 305–329). SAGE Publications, Inc. https://doi.org/10.4135/9781412986274.n12

Roh, T.-H., & Xiao, S. (2024). Extending the research agenda for supply chain management in the age of disruption: The multifaceted role and implications of dynamic capabilities. *Journal of General Management*, 50(1). https://doi.org/10.1177/03063070241272373

Saeed, A., & Shoukat, R. (2023). Lean, green, and agile supply chain practices. In M. Waqas, S. Khan, & A. Al-Amin (Eds.), *Emerging Trends in Sustainable Supply Chain Management and Green Logistics* (pp. 121–142). IGI Global Scientific Publishing.

Samik, H. Y. A., & Al-Khashman, A. M. (2024). The impact of dynamic capabilities on supply chain performance at Jordanian pharmaceutical manufacturing companies. *International Journal of Academic Research in Business and Social Sciences*, 14(8). https://doi.org/10.6007/IJARBSS/v14-i8/22583

Santos, G. P. R. dos, Silva, E. V. da, Corrêa, A. de C., Silva Filho, A. B. da, Imbiriba, K. G., Carr, R. F., Batista, E. D., Lima, J. N. P. de, Souza, R. B. R. A., Santos, F. M. T. dos, Silva, G. dos S., & Coimbra, M. P. C. (2024). Collaborative logistics systems for sustainability: Exploring opportunities and limitations in reducing environmental footprints in supply chains. *Contribuciones a Las Ciencias Sociales*, 17(12), 1–24. https://doi.org/10.55905/revconv.17n.12-388

Schianchia, M. (2023). IoT-based framework for agile supply chain in FMCG industries – A solution for (post-)pandemic times. *International Journal of Innovation in Engineering*, 3(1), 63–70. Retrieved from http://www.ijie.ir

Shafaghizadeh, S., & Sajadi, S. M. (2023). Ranking agility factors to reliably sustain a green industrial supply chain using the fuzzy analytic network process and ordinal priority approach. *Sustainability*, 15, 16005. https://doi.org/10.3390/su152216005

Singh, R. K., & Mathiyazhagan, K. (2024). Integrating flexibility and sustainability in supply chains: A practice-based view on leadership, knowledge and digital quality management practices. *The International Journal of Logistics Management*, 35(6), 1961–1980. https://doi.org/10.1108/IJLM-02-2024-0082

Tabesh, A. R., Chowdhury, M. H., Quaddus, M. A., Sianaki, O. A., & Khan, E. (2024). The influence of agility on green purchasing practices: The role of supplier governance and supplier relationships. *Journal of Business & Industrial Marketing*. https://doi.org/10.1108/jbim-04-2023-0205

Tang, C., & Tomlin, B. (2008). The power of flexibility for mitigating supply chain risks. *Internal Journal of Production Economics*, 116, 12–27. https://doi.org/10.1016/j.ijpe.2008.07.008

Teece, D. J., Pisano, G., & Shuen, A. (1997). Dynamic capabilities and strategic management. *Strategic Management Journal*, 18(7), 509–533. https://doi.org/10.1002/(SICI)1097-0266(199708)18:7<509::AID-SMJ882>3.0.CO;2-Z

Thomas, J., & Harden, A. (2008). Methods for the thematic synthesis of qualitative research in systematic reviews. *BMC Medical Research Methodology*, 8, 45, https://doi.org/10.1186/1471-2288-8-45

Towers, N., & Burnes, B. (2008). A composite framework of supply chain management and enterprise planning for small and medium-sized manufacturing enterprises. *Supply Chain Management: An International Journal*, 13(5), 349–355. https://doi.org/10.1108/13598540810894933

Tuliakova, A., Карінцева, O., & Тарасенко, C. (2024). Dynamic capabilities as a success-factor of the long-term business alliance: The case of Siemens and Atos. Підприємництво та Інновації *(Entrepreneurship and Innovation)*. https://doi.org/10.32782/2415-3583/32.16

Uddin, M. H., Razzak, M. R., & Rahman, A. A. (2023). Sustainable supply chain management practices, dynamic capabilities, and competitive advantage: Evidence from Bangladesh ready-made garments industry. *Business Strategy & Development*, 1(1), 1–13. https://doi.org/10.1002/bsd2.232

Vozdvyzhenskyi, V., & Titova, M. M. (2024). Competitiveness of the enterprise: Theoretical aspects of formation and practical tools of implementation. *Ekonomìčnij vìsnik Dnìprovs'kogo deržavnogo tehnìčnogo unìversitetu (Economic Bulletin of the Dniprovsk State Technical University)*. https://doi.org/10.31319/2709-2879.2024iss2(9).318854pp58-65

Weerawardena, J., & Mavondo, F. T. (2011). Capabilities, innovation and competitive advantage. *Industrial Marketing Management*, 40, 1220–1223. https://doi.org/10.1016/j.indmarman.2011.10.012

Whitten, G.D., Green Jr., K.W., & Zelbst, P.J. (2012). Triple-A supply chain performance. *International Journal of Operations & Production Management*, 32(1), 28–48. https://doi.org/10.1108/01443571211195727

Wilden, R., Gudergan, S. P., Nielsen, B. B., & Lings, I. (2013). Dynamic capabilities and performance: Strategy. *Structure and Environment, Long Range Planning*, 46, 72–96. https://doi.org/10.1016/j.lrp.2012.12.001

Wu, F., Yeniyurt, S., Kim, D., & Cavusgil, S. T. (2006). The impact of information technology on supply chain capabilities and firm performance: A resource-based view, *Industrial Marketing Management* 35, 493–504. https://doi.org/10.1016/j.indmarman.2005.05.003

Yuthas, K., Sarason, Y., & Aziz, A. (2021). Strategic value creation through enterprise blockchain. *The Journal of the British Blockchain Association*, 4(1). https://doi.org/10.31585/jbba-4-1-(7)2021

Section II

CSR, Governance, and Social Marketing

4 Stakeholder Influences on Sustainable CSR Strategies for Supply Chain Networks

Benjamin Silas Bvepfepfe

4.1 INTRODUCTION

There is evidence that suggests the evolution of supply chain management in relation to responding to economic, social, and environmental factors in the supply chain operational environments. For chief supply chain officers (CSCO), the notions of corporate social responsibility (CSR) have become key pillars for the development and execution of sustainable supply chain strategies. There is evidence that suggests that embedding CSR within supply chain strategies can result in sustainable benefits for stakeholders. To avoid CSR programmes and strategies being considered unsustainable, or mere 'window dressing' or 'superfluous', stakeholder engagement is therefore crucial (Freeman and Velamuri, 2023).

4.2 PROBLEM STATEMENT

CSR has been used interchangeably with sustainability, such that potentially the adoption of initiatives is likely to be varied across global supply chains. Potentially, some problems may arise as to the nature of CSR due to the geographical location of supply chain operations. It becomes essential for organisations and their supply chain partners to have a framework that can be adopted across the supply chain in order to be responsive to the expectations and demands of key stakeholders.

4.2.1 Research Question

The key research question relates to the motivations and drivers of CSR initiatives for organisations and their global supply chain partners. This question is further split to identify the key stakeholders and how they are considered when organisations develop plans for CSR initiatives.

DOI: 10.1201/9781003770008-6

4.2.2 Research Gap

Whilst research appears to cover CSR definitions and its impact on corporate performance, there is a need to explore the transferability of CSR strategies across organisational boundaries, especially for globalised supply chain operations.

4.2.3 Rationale

It can be argued that the materiality of CSR practices is considered responsive when these are derived from an effective stakeholder dialogue method. The challenge is identifying the materiality of CSR issues in order to prioritise the same from a multiplicity of expectations and demands. For example, CSR strategies and practices must therefore be adopted throughout an organisation's supply chain network to be considered appropriate for the stakeholders' demands and expectations.

4.2.4 Objectives

The chapter investigates key stakeholder influences on CSR practices for global supply chain networks.

The key objectives are:

- Analyse stakeholder theory and its influence on organisational CSR practices
- Investigate the extend that stakeholder considerations are considered in the CSR materiality of organisations
- Construct a stakeholder model for developing and implementing CSR strategies for supply chain operations

4.2.5 Scope

The scope of this chapter covers key theories of stakeholders and how these have influenced CSR decisions within the sampled organisations.

4.2.6 Limitations

The time frame for data collection could be increased to about five years. This would enhance interpretations of results and the ensuing assumptions regarding the stakeholder influences on CSR prioritisation.

4.2.7 Structure of Chapter

This chapter is structured around an introduction, followed by a detailed literature review and methodology section. The next sections cover the discussion of key results and the chapter's conclusion.

4.3 LITERATURE REVIEW

4.3.1 Introduction

Supply chain operations have become the focus of much attention not only from organisational perspectives but also from stakeholder perspectives. Govindan et al. (2020) highlighted that it is essential for businesses and their supply chain partners to measure the sustainability of their supply chain operations. This literature review explores the tenets of supply chain sustainability, the stakeholder influences on sustainability, and the relationship between sustainability and CSR practices.

4.3.2 From Supply Chain Management to Sustainable Supply Chain Management

Managing supply chains gained prominence from as early as 1980s, as businesses recognised that competitive advantage can be maximised by managing and controlling the organisation's operations and resources, beyond the organisation's boundaries. Previous studies report several benefits accruing from supply chain management, including efficiencies in inventory (e.g., reduction in unnecessary stock holding, storage costs) and improved delivery lead times. Baah et al. (2022) further suggested that enhanced collaboration between supply chain partners sustains effective long-term relationships amongst upstream and downstream supply chain partners, thereby influencing overall performance. Sustainability appears to reflect the gains that CSR has made for business, providing platforms for initiatives like recycling chains and other social sustainability issues like improvement to working conditions and reduction of environmental negative impacts. In their study on supply chain sustainability within the fast fashion industry supply chains, Wren (2022) noted that sustainability initiatives failed due to a lack of collaborative incentives between upstream and downstream partners.

4.3.3 CSR and Supply Chain Sustainability

CSR has connotations that vary with interpretations as far as the development and deployment of organisational initiatives (Sánchez-Teba et al., 2021). The notions of CSR have developed to become important strategic issues for business organisations. There are various forces that pushed and, in some cases, encouraged businesses to engage in CSR-related initiatives. For example, it is now evident that supply chain operations are expected to report to specific key stakeholders, how they are responding to these forces in various channels of communication. Whilst there is no agreed definition for CSR, contributors like Arco-Castro et al. (2020) consider CSR as a strategic management approach that can be adopted to gain a competitive advantage within globalised supply chains. They went further to suggest that supply chain partners must consider stakeholder expectations when developing sustainable strategies.

Trace of CSR conceptualisation is summarised hereunder:

- Philanthropic CSR initiatives aim to provide an organisation with advantages in terms of reputation and social capital.
- Profit and shareholder value maximisation, emphasising that the only responsibility of an organisation is profit and shareholder value maximisation.
- Stakeholder theory has gained ground as a complementary framework to explore CSR, especially from a social issue perspective. According to Awa et al. (2024), the stakeholders' approach to CSR proactively strengthens the interactions between various key players, especially in relation to sustainability and environmental issues.
- The notion that CSR is best adopted through a systematic integration within corporate strategy has gained ground. This notion is based on arguments that the materiality of CSR issues and their implementation should be embedded into the organisation's strategies for sustainable responsiveness (Fatima and Elbanna, 2023).

There are proponents led by Friedman (1970) who argue that the CSR concept violates the shareholder obligations of organisations. On the other hand, others still argue that organisations use CSR as a cover-up to corporate ills and mislead stakeholders to protect the organisation's brand or reputation. For example, an organisation whose operations have damaging effects on the environment would engage in CSR initiatives only to window dress the negative impact its operations have on the environment. Similarly, an organisation that manufactures products that are harmful to public health, for example, tobacco or alcohol, would build schools or sponsor sporting events as part of its community service. The CSR concept and stakeholder theory, although distinct, create a platform and overlap that remove some of the contentions on the role of businesses in society. Viewing CSR as those activities directed towards society is a limited view of CSR. Supply chain strategies must consider embedding CSR initiatives derived from expectations of key stakeholders. In this view, it is the stakeholders who can evaluate the responsiveness and materiality of CSR initiatives. The question is which stakeholders are capable and have the capacity to evaluate CSR initiatives meaningfully. There are arguments from those proponents of the free market that not all stakeholders have a legitimate interest.

4.3.4 Stakeholder Perspectives

The influences of stakeholders on the formulation of sustainable CSR performance across supply chains have been explored by researchers and other contributors (Jabbour et al., 2020). A stakeholder is defined in terms of any individual or group that has the capacity to affect or be affected by the organisation's efforts towards the achievement of the desired goals (Freeman, 2010). It is clear from recent contributions that the stakeholder concept now provides much more scope for CSR discourse in relation to the materialisation and prioritisation of CSR issues. For example, stakeholders can be grouped in relation to their level of interest and how they are impacted by the organisation's and its supply chain operations.

4.3.5 Stakeholder Theory and CSR

When the stakeholder theory is applied for assessing CSR, as alluded to by Clarkson (1995), social performance can be evaluated in relation to stakeholder demands or expectations. This implies that responsiveness and achievement of CSR objectives are determined by the level of engagement between the organisation and its stakeholders. In globalised supply chain operations, the expectations and demands of diverse stakeholders present organisations with challenges which specific issues and interests to prioritise. For example, supply chain network partners, upstream and downstream, would endeavour to offload CSR demands to the other. Clarkson (1995) classified two groups of stakeholders, that is, primary and secondary stakeholder groups. The primary group comprises those stakeholders who are considered critical and key for the organisation's survival, for example, shareholders, workers, consumers, sellers, the government, and society. In addition, this primary group has direct reciprocal and transactional relationships with the organisation. It is possible to assume that CSR initiatives directed towards the primary group of stakeholders, for example, employees or clients, are likely to enhance trust and improve productivity.

A secondary stakeholder has indirect influence on an organisation and its supply chain partners. These stakeholders can still influence or disrupt supply chain operations. This secondary stakeholder group assumes a key role in evaluating CSR activities and organisational social responsiveness of management decision-making. This group exercises their influences through primary stakeholders, especially in aspects of ethical and reputational CSR matters. Others have proposed a more complicated classification of stakeholders into six main categories: regulators, controllers, partners, passive stakeholders, dependent stakeholders, and non-stakeholders. The classification is determined by the relationship, relevance, influence, and participation of stakeholders as the key variables between the organisation and its stakeholders.

It would follow that the most effective approach to social responsiveness is to adopt the stakeholder theory in developing CSR programmes. This view is supported by Freeman and Velamuri (2023) that CSR goals should focus on enhancing value to key stakeholders. In these multifaceted stakeholder expectations, supply chain networks struggle to balance these expectations and overall corporate objectives. Therefore, the purpose of stakeholder management in CSR strategic planning is to identify stakeholder groups and their conflicts, so that these demands and interests are navigated through. This stakeholder–CSR perspective has widened the significance of societal concerns and the resultant shape that CSR strategies take within given operational supply chain environments.

Contributors like Freeman (2010) suggested that any framework seeking to manage stakeholders should begin with a clear definition, in order to identify who the stakeholders are, in pursuit of its mission. According to Freeman, this approach enables the corporation to gain insights into the interests and motivational influences of their relationship with certain stakeholders. It is clear that the lack of a clear process of identifying and mapping stakeholders would lead to challenges in strategy formulation and implementation. So, the framework becomes a central issue here for successful supply chain operations and sustainable CSR strategies.

There are various industry sectors where stakeholder theory has been adopted to formulate CSR-related initiatives, for example, mining, oil and gas, construction, and engineering and manufacturing processes.

4.3.6 Application of Stakeholder Theory to CSR

The stakeholder theory is considered appropriate in advancing the CSR practices within organisations and their supply chain operations. Other contributors have advanced the notion that there can be no sustainable CSR initiatives without stakeholder engagement and that the quality of stakeholder relationships with the organisation determines CSR responsiveness (Freeman and Velamuri, 2023). The instrumental stakeholder theory can give an organisation some competitive edge within the global marketplace. As CSR is a stakeholder-oriented concept, this assumes that organisations should have dialogue methods with key stakeholders to determine the materiality of CSR issues.

It is agreed that when organisations engage in CSR activities, they portray an impression of doing good, for example, community-related projects that aim at improving the well-being of society. Some organisations have policies that encourage their employees to volunteer for supporting local community-related projects. There are arguments that some of these CSR initiatives are used by organisations and their supply chain partners to cushion against seemingly bad global supply chain operations. For example, an organisation manufacturing tobacco that is considered harmful to the health of the community may decide to build schools and even health centres in some locations of its global supply chain operations.

There is scope for organisations and their supply chain partners to maximise stakeholder relationships, which will improve the levels of trust and loyalty with key stakeholders, for example, suppliers, employees, and communities. Consider an organisation that engages in initiatives towards environmental improvements; this action will certainly increase its moral reputation with some key stakeholders, especially the secondary group of stakeholders. Mapping stakeholders to identify the key attributes and the nature of their expectations and demands provides greater insights for organisations when developing CSR programmes or strategies.

It can be argued that stakeholder mapping is essential for designing responsive CSR programmes that should meet the multiplicity of stakeholder expectations within prevailing contexts. Organisations should assess the contextual relationships within their specific operational environment to identify demands and pressure from key stakeholders. The study by Thanetsunthorn and Wuthisatian (2025) revealed strong correlations between institutional environments and CSR uptakes. For example, governments in emerging markets have provided incentives for organisations to take up more societal initiatives.

For example, organisations and their supply chain partners will list all individuals, groups of individuals, organisations, departments, entities, and networks as potential stakeholders. This pragmatic and broad view approach to strategic analysis assumes that organisations have multiple stakeholder relationships with everyone who can be affected or can affect an organisation's strategic performance. The challenge faced by organisations and their supply chain networks is deciding and defining the materiality of CSR issues to prioritise, given the multiplicity of stakeholders.

The narrow view attempts to classify stakeholders based on their direct relevance to the organisation's business objectives. This view differs from the broad view, as it groups stakeholders by their closeness to business operations, for example, between primary and secondary stakeholders (Clarkson, 1995). This could be derived from assessing the stakeholder position in relation to their perceived value, adding to the objectives of the organisation and its supply chain. Stakeholder mapping assumes that key stakeholders within given contexts or situations will influence decisions that determine how the organisation prioritises CSR initiatives.

4.4 METHODOLOGY

This chapter aims to explore in more detail and to delve deeper into the meanings and the stakeholder perspective in CSR across global supply chain networks. The case study approach provides an appropriate contextual setting for deeper insights into the adoption and practices of CSR across supply chain operations (Yin, 2003). The case study allowed for both qualitative and quantitative data to be analysed from sample organisations.

4.4.1 Sample and Data Collection

A purposeful sample of organisations considered to be leading in CSR in the United Kingdom, South Africa were selected for this inquiry. Business in the Community (BiTC) UK, who have a membership ranking for organisations that engage in CSR initiative, was considered to be appropriate for organisation samples for data collection. In South Africa, the Johannesburg Stock Exchange (JSE) has an index, the Social Responsibility Index (SRI), that classifies organisations according to their published CSR initiatives. From these data sources, sample organisations and key people were identified for the purpose of data collection and administering semi-structured questionnaires regarding motivations, key stakeholders, and CSR issues. Corporate publications, like annual reports and websites, were used to identify themes related to CSR initiatives.

The researcher used thematic analysis for grouping of words in order to ascertain assumptions and meanings of underlying themes or concepts. For example, a step-by step process identified words and key themes that were highlighted to quantify CSR definitions and key issues in the sample organisations. In the analysis, the three pillars of sustainability – economic, social, and environmental issues – were applied for grouping of CSR-related issues. Mahmood et al. (2021) would call these as first-order themes. Then four key CSR subthemes or 'second-order themes', namely, workplace, market place, community, and environmental, were selected from the dimensions and are considered to be consistent with business operations and can be easily aligned to the sustainability first-order themes above. Sustainability researchers and other industry associations like BiTC UK, JSE, and SRI have adopted similar themes for evaluating CSR responsiveness. Third-order schemes provide brief explanations of the dimensions above, while the fourth-order scheme provides some examples to make reading and understanding easier.

Content analysis (CA) was adopted to assess the context of messages and emerging themes, in order to make a deduction to CSR theory. As annual reports are considered a dialogue vehicle between the organisation and its stakeholders, they are considered important data sources for the research. Furthermore, it is also clear that most organisations are now using these reports to disseminate their social, economic, and environmental initiatives.

4.5 DISCUSSION OF RESULTS

Based on the results, the organisational CSR motivations were grouped into six dimensions. These dimensions fit into the three basic types of motives why organisations engage in CSR, namely, intrinsic motives, extrinsic motives, and stakeholder motives. In this case, extrinsic motives are evident in performance-driven and risk management dimensions. These CSR initiatives are focused more on the organisation's strategic output and are considered self-serving CSR. However, the intrinsic motives are the value-driven and community-related motivations. Organisations engage in CSR initiatives as a public-serving, altruistic motive. There appears to be an interest in societal well-being, thereby creating an improved corporate image. Lastly, the stakeholder motive is catered for in the stakeholder-driven and legal compliance dimensions.

The results show that organisations report their CSR initiatives as part of the annual reporting systems in a variety of ways. The results provided evidence of two strands of reporting emerging, that is, dedicated CSR reports and CSR reporting as part of the main annual report. Dedicated reports are produced as separate and standalone CSR reports that appear to provide insights into the organisation's CSR activities. This appears to respond to criticisms that CSR reporting lacked sufficient completeness in content. Benchmarking appears to be a key feature for some sample organisations reports, and this is evidence of attempts towards harmonisation of reporting with industry standards. There are suggestions that supply chains will benefit from harmonised sustainability reporting across the whole pipeline as this may fill the gap from weaker governance institutional environments (Scuotto et al., 2022). In fact, information flows are a key driver for effective supply chain performance, and in this particular case, global supply chains will benefit from standardised sustainability reporting.

4.5.1 Comparative Analysis

The results suggest awareness of the value of CSR initiatives and reporting in sample organisations, aligning with suggestions that more companies are publishing and reporting on their CSR initiatives than before (Nejati and Shafaei, 2023). There are some differences in the disclosure of CSR initiatives in relation to report size and content coverage. Some organisations prefer to have separate and stand-alone reports for their sustainability efforts, whilst others confine CSR details within the main annual reports.

This supports the notion that CSR issues may not necessarily reflect similar attributes across the global supply chain. The CSR disclosure in the same organisations' reports is more substantive in nature than symbolic. As the organisations are from the extractive and global retail industries, this would support the notion that such organisations tend to disclose more of their activities in response to the legal compliance requirements. However, the stand-alone CSR reports are extensive and appear to be ingrained into the organisation's strategy and overall engagement with stakeholders.

4.5.2 Stakeholder Influences on CSR Initiatives

The results revealed a group of key stakeholders that appear to be targeted in the communication strategies for organisations and their supply chain networks. There are some connotations in the communications strategies that some organisation only engages with stakeholders as part of public relations, whilst in other situations, organisations communicate with their stakeholders in order to determine and respond to specific CSR issues. As revealed in the motivations and the materiality of issues, organisations have explicitly stated their social accountability in relation to societal, sustainability, market-related, and compliance-driven issues. The importance of stakeholders is also demonstrated by the dialogue methods that are in place to engage with selected stakeholder groups.

It is evident that stakeholders have been identified and mapped; however, the grouping lacks crucial strategic value so as to assign an appropriate classification for strategic purposes. Looking across the sample organisation, it would appear that similar stakeholders are identified for the same industries, especially those in high-visibility operations.

4.5.3 CSR Responsiveness Through Stakeholder Engagement

The identification of CSR issues for sample organisations appears to be aligned with the particular needs of stakeholders. This would resonate with Nejati and Shafaei (2023) in the discourse that corporate objectives will be subjected to scrutiny by the organisation's stakeholders. This would imply that those organisations that facilitate stakeholder engagement within their supply chain networks are likely to have higher levels of CSR responsiveness.

The stakeholder theory forms a foundational basis for organisations and their supply chain operations to develop meaningful CSR initiatives, rather than mere window dressing or marketing attempts. It is also important to note that variations are likely to emerge across organisations and their network partners due to environmental contexts. This will entail different priorities and orientations for CSR strategies.

4.6 CONCLUSION

Organisations appear to engage stakeholders to reflect on CSR initiatives across their supply chain networks. Whilst CSR strategies will be varied across supply chain networks, it is assumed that organisations appear to embed CSR as a value-adding activity rather than a cost. Stakeholders will continue to be identified for purposes of developing sustainable strategies. It is also evident that institutional settings or country contexts will influence the level or nature of CSR initiatives. This is an important part of global supply chains whose operations cross regional boundaries. The identification of stakeholders should go a step further in setting out clearly what the specific expectations and demands from these stakeholders are. This would ensure that initiatives are more aligned and suitable to these expectations.

4.6.1 Limitations

The time frame for data collection could be increased to about five years. This would enhance interpretations of results and the ensuing assumptions regarding the stakeholder influences on CSR prioritisation. Expansion of the data collection base to include other regional supply chains would also enhance the global perspective on CSR.

4.6.2 Contributions/Novelties

This inquiry has reinforced Mendelow's (1991) model by incorporating the contextual factors into the formulation of CSR strategies. For example, the power/interest mapping process of stakeholders can be effective within a given context and institutional environments. The assumption here is that the more formalised the institutions, the higher the level of CSR initiatives. This is an area that needs further research and development.

REFERENCES

Arco-Castro, M. L., Lopez-Pérez, M. V., Rodriguez-Gomez, S., & Garde-Sánchez, R. (2020). Do stakeholders modulate philanthropic strategy? Corporate philanthropy as stakeholders' engagement. *Sustainability*, *12*(18), 7242.

Awa, H. O., Etim, W., & Ogbonda, E. (2024). Stakeholders, stakeholder theory and corporate social responsibility (CSR). *International Journal of Corporate Social Responsibility*, *9*(1), 11.

Baah, C., Acquah, I. S. K., & Ofori, D. (2022). Exploring the influence of supply chain collaboration on supply chain visibility, stakeholder trust, environmental and financial performances: A partial least square approach. *Benchmarking: An International Journal*, *29*(1), 172–193.

Clarkson, M. E. (1995). A stakeholder framework for analyzing and evaluating corporate social performance. *Academy of Management Review*, *20*(1), 92–117.

Fatima, T., & Elbanna, S. (2023). Corporate social responsibility (CSR) implementation: A review and a research agenda towards an integrative framework. *Journal of Business Ethics*, *183*(1), 105–121.

Freeman, R. E. (2010). *Strategic Management: A Stakeholder Approach*. Cambridge University Press.

Freeman, R. E., & Velamuri, S. R. (2023). A new approach to CSR: Company stakeholder responsibility. In S. D. Dmytriyev, & R. Edward Freeman (eds.), *R. Edward Freeman's Selected Works on Stakeholder Theory and Business Ethics* (pp. 251–264). Cham: Springer International Publishing.

Friedman, M. (1970). The social responsibility of business. *The New York Times Magazine*.

Govindan, K., Rajeev, A., Padhi, S. S., & Pati, R. K. (2020). Supply chain sustainability and performance of firms: A meta-analysis of the literature. *Transportation Research Part E: Logistics and Transportation Review*, *137*, 101923.

Jabbour, C. J. C., Seuring, S., de Sousa Jabbour, A. B. L., Jugend, D., Fiorini, P. D. C., Latan, H., & Izeppi, W. C. (2020). Stakeholders, innovative business models for the circular economy and sustainable performance of firms in an emerging economy facing institutional voids. *Journal of Environmental Management*, *264*, 110416.

Mahmood, A., Naveed, R. T., Ahmad, N., Scholz, M., Khalique, M., & Adnan, M. (2021). Unleashing the barriers to CSR implementation in the same sector of a developing economy: A thematic analysis approach. *Sustainability*, *13*(22), 12710.

Mendelow, A. (1991). *Stakeholder mapping*. In *Proceedings of the Second International Conference on Information Systems* (pp. 407–418). Cambridge, MA, Great Britain.

Nejati, M., & Shafaei, A. (2023). Why do employees respond differently to corporate social responsibility? A study of substantive and symbolic corporate social responsibility. *Corporate Social Responsibility and Environmental Management*, *30*(4), 2066–2080.

Sánchez-Teba, E. M., Benítez-Márquez, M. D., Bermúdez-González, G., & Luna-Pereira, M. D. M. (2021). Mapping the knowledge of CSR and sustainability. *Sustainability*, *13*(18), 10106.

Scuotto, V., Chin, T., Pezzi, A., & Pironti, M. (2022). CSR best practices for global multi-tier sustainable supply chain integration of Chinese MNEs. *Corporate Social Responsibility and Environmental Management*, *29*(6), 2038–2052.

Thanetsunthorn, N., & Wuthisatian, R. (2025). Corporate social responsibility in emerging markets: The role of reciprocity in business-government relations. *Social Responsibility Journal*, *21*(3), 520–548.

Wren, B. (2022). Sustainable supply chain management in the fast fashion Industry: A comparative study of current efforts and best practices to address the climate crisis. *Cleaner Logistics and Supply Chain*, *4*, 100032.

Yin, R. K. (2003). Designing case studies. *Qualitative Research Methods*, *5*(14), 359–386.

5 Logistics Performance and Environmental Sustainability

Mapping Research Trends and Future Directions Through Systematic Review

Khaled Ben Abdallah, Firas Ben Romdhane, and Fedy Ouni

5.1 INTRODUCTION

The efficiency of logistics systems is essential for strengthening economic productivity, facilitating cross-border trade, and sustaining industrial growth (Khan et al., 2019; Rashidi & Cullinane, 2019). Despite its economic role, the logistics sector is an important source of emissions. Additionally, the sector's considerable carbon footprint and energy consumption pose challenges to align with the increasing demand for sustainability (Ali et al., 2021; Lin et al., 2021).

Given the growing importance of harmonizing logistics, economic growth, and environmental sustainability, this subject has attracted considerable academic interest (Khan et al., 2017; Ouni et al., 2023). Several critical gaps and contradictions persist in the literature. First, researchers employ diverse indicators, time periods, and analytical approaches without standardized frameworks. Second, there is a noticeable geographical concentration on developed economies, which restricts insights into developing contexts where logistics systems evolve under distinct economic, institutional, and regulatory conditions. Third, most studies offer cross-sectional perspectives rather than longitudinal analyses. Finally, literature lacks a thorough synthesis of policy interventions and their effectiveness in achieving sustainable logistical transitions, limiting practical guidance for decision-makers seeking evidence-based solutions that balance economic and environmental goals.

Major research gaps exist. First, there is a limited integration of environmental and economic consequences in logistics research. Second, a limited multicountry comparative analysis is focused on Asia and the Organisation for Economic Co-operation and Development (OECD) countries. The standardized measures are

DOI: 10.1201/9781003770008-7

needed to assess green logistics performance. Finally, the mediating role of institutional quality and governance is underexplored.

To respond to these challenges, this study employs a systematic literature review, analyzing 65 articles published between 2010 and 2023, to explore how logistics performance interacts with economic and environmental factors.

This chapter is organized into four sections. Section 5.2 details the research methodology. Section 5.3 presents the findings from the descriptive analysis. Section 5.4 discusses the evolution of logistics policies. Finally, the study concludes with implications for policy.

5.2 METHODOLOGY

5.2.1 Search Strategy

This literature review focuses on several key electronic databases (Scopus and Web of Science). As highlighted in Table 5.1, the search strategy was organized around three conceptual dimensions: logistics performance, economic conditions, and environmental factors.

5.2.2 Inclusion and Exclusion Criteria

Studies were retained if they were published in peer-reviewed journals in English. They were excluded if they consisted of conference abstracts, editorials, book reviews, or non-scholarly reports, or if logistics performance was not a central analytical component.

5.2.3 Study Selection Process

The database searches yielded over 1,040 articles. The selection process involved five stages: filtering by English language and time (446 exclusions), removing duplicates and irrelevant titles/abstracts (491 exclusions), discarding unavailable full texts (23 exclusions), conducting full-text reviews of 80 articles (31 exclusions), and examining references of the remaining 49 studies to identify additional works. Ultimately, only 16 articles met the inclusion criteria for the final review (Figure 5.1).

TABLE 5.1
Representative Search Terms

Dimension	Keywords
Logistics performance	Logistics performance, logistics efficiency, supply chain performance, logistics performance index, LPI
Economic indicators	Economic growth, GDP, economic development, trade, foreign direct investment
Environmental factors	Environmental degradation, environmental sustainability, green logistics, sustainable development, carbon emissions, CO_2 emissions

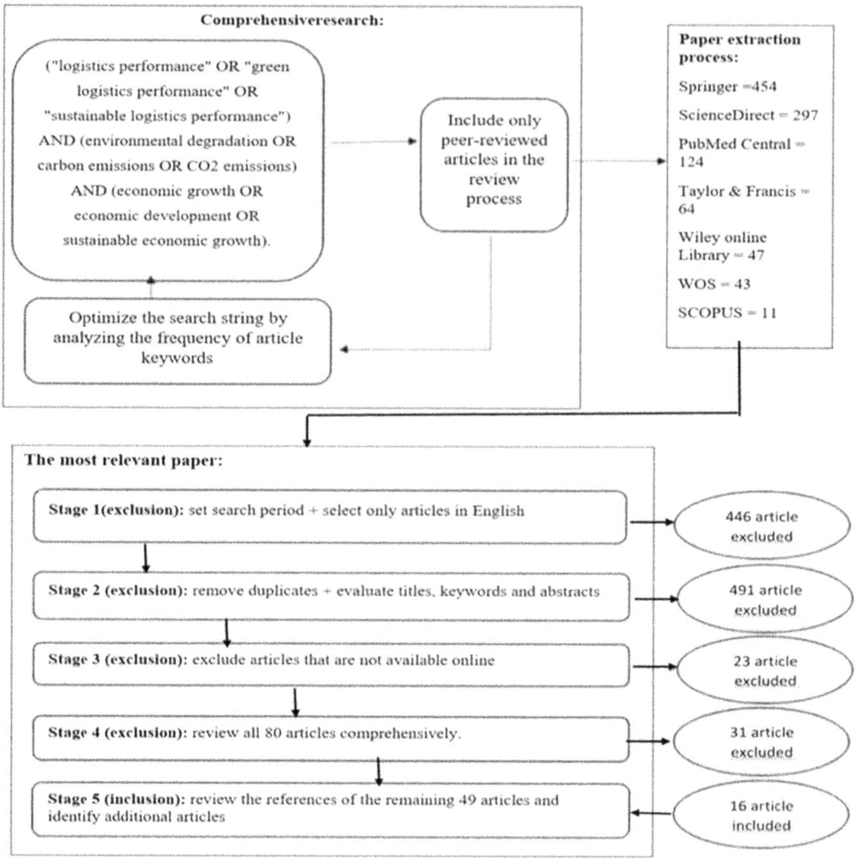

FIGURE 5.1 Process of SLR.

5.3 RESULTS AND DISCUSSIONS

5.3.1 Publication by Year

Figure 5.2 represents a substantial evidence base spanning 14 years of research (2010–2023) and demonstrates the growing academic interest in the intersection of logistics, economics, and environmental sustainability.

As shown in Figure 5.2, there is a noticeable upward trend in publications. In all, 72% of publications were published between 2017 and 2023. Twelve publications peaked in 2022, maybe due to increased climate change and sustainable development goals.

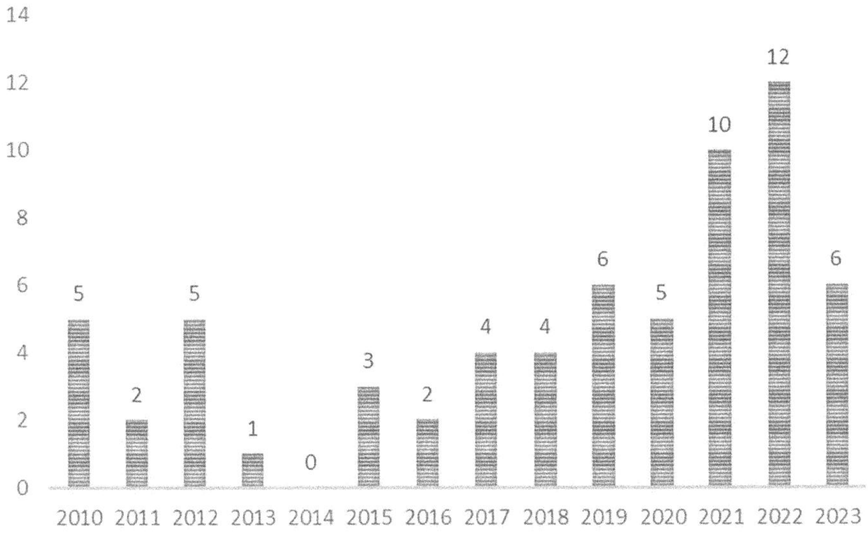

FIGURE 5.2 Yearly publication (2010–2023).

5.3.2 Journal Distribution and Publication Patterns

The 65 selected articles were published across 37 different journals, indicating the interdisciplinary nature of research in this field. Table 5.2 highlights 11 journals that have published at least two articles, along with their H-index as determined by the Scimago classification. The highest number of publications is in *Environmental Science and Pollution Research* journal (10 articles), followed by the *Journal of Cleaner Production* with six articles, and *Sustainable Development* with five articles.

TABLE 5.2
Distribution of Articles by Journal

Journal	Number of Papers	H-Index
Environmental Science and Pollution Research	10	154
Journal of Cleaner Production	6	268
Sustainable Development	5	81
Economic Research	3	43
Sustainability	3	136
Environment, Development and Sustainability	2	72
Frontiers in Environmental Science	2	61
Heliyon	2	69
Journal of Environmental Economics and Management	2	141
PLoS One	2	404
Transportation Research Part A: Policy and Practice	2	153
Total	**65**	

5.3.3 Geographical Distribution and Regional Research Patterns

Table 5.3 summarizes the geographical coverage of studies employing multicountry analyses. The distribution of publications shows a pronounced concentration on Asian contexts.

Among single-country studies (Table 5.4), China clearly dominates the literature, with 11 articles devoted exclusively to its national case, making it the most frequently examined country in the sample. This level of attention contrasts sharply with other regions and reflects China's central role in global logistics networks and sustainability-related research. Multicountry studies represent a sizeable share of the reviewed

TABLE 5.3
Geographical Distribution of Studies Based on Multicountry Analyses

Reference(s)	Region Studied	No. of Articles
Gu et al. (2022), Bansal et al. (2021), Akram et al. (2023), Liu et al. (2018), Khan et al. (2019), Liu et al. (2023), and Suki et al. (2021)	Asian countries	8
Aydin et al. (2023), Li et al. (2021), Mohsin et al. (2022a, 2022b), and An et al. (2021)	BRI countries	4
Aichele and Felbermayr (2012), Mujtaba and Shahzad (2021), Hayaloğlu (2015), and Khan et al. (2022)	OECD countries	4
Menyah and Wolde-Rufael (2010), Odusola and Abidoye (2015), and Barrios et al. (2010)	African countries	3
Borhan et al. (2012) and Khan (2019)	ASEAN countries	2
Zaman and Shamsuddin (2017) and Khan et al. (2021)	European countries	2
Mariano et al. (2017)	104 countries	1
Lu et al. (2019)	112 countries	1
Jia et al. (2022)	119 countries	1
Dell et al. (2012)	125 countries	1
Jomthanachai et al. (2022)	134 countries	1
Khan et al. (2016)	115 globally ranked logistics countries	1
Khan et al. (2018)	19 developed countries	1
Wan et al. (2022)	22 emerging countries	1
Kim and Min (2011)	23 countries	1
Saidi et al. (2020)	46 developing countries	1
Coto-Millán et al. (2013)	Major world countries	1
Li et al. (2021)	South American countries	1
Karaduman et al. (2020)	12 Balkan countries	1
Amirnejad et al. (2021)	MENA countries	1
Fan et al. (2022)	15 RCEP countries	1
Khan et al. (2019)	SAARC countries	1
Yu et al. (2018)	43 countries worldwide	1

TABLE 5.4
Geographical Distribution of Studies Based on Single-Country Analyses

Reference(s)	Country	No. of Articles
Chu (2012), Godil et al. (2021), Peng et al. (2020), Ding et al. (2023), Khan et al. (2022), He et al. (2017), Jiang et al. (2020), Yunlin (2023), Li et al. (2021), Wang and Wu (2023), and Hao et al. (2021)	China	11
Bölük and Mert (2015), Büyüközkan and Çifçi (2011), and Göçer et al. (2022)	Turkey	3
Bello and Abimbola (2010) and Kusi-Sarpong (2019)	Nigeria	2
Islam et al. (2021) and Raggad (2018)	Saudi Arabia	2
Prajogo et al. (2012)	Australia	1
He and Richard (2010)	Canada	1
Alvarado and Toledo (2017)	Ecuador	1
Ghosh (2010)	India	1
Larson (2021)	Switzerland	1
Khan and Qianli (2017)	UK	1
Moreira and Rodrigues (2023)	Brazil	1

work and span a variety of regional groupings, including Asian countries, OECD economies, Belt and Road Initiative (BRI) countries, African countries, Association of Southeast Asian Nations (ASEAN) members, and European regions. Several contributions adopt a global scope, drawing on large cross-country datasets covering between 19 and more than 130 countries. North Africa and the broader Middle East and North Africa (MENA) region are scarcely represented.

5.3.4 Methodological Approaches and Analytical Techniques

Figure 5.3 indicates that the methodological analysis demonstrates significant variability in analytical approaches, highlighting the complexity of the research problems and the interdisciplinary character of the area.

The findings of Ali et al. (2023) reveal that environmental pollution declines with increased use of renewable energy. In sub-Saharan Africa, Barrios et al. (2010) discuss the impact of rainfall trends on economic growth performance compared to other developing nations. New transnational climate data integrated into the empirical framework of economic growth demonstrate that rainfall has a direct correlation with weaker economic growth.

Ghosh (2010) examines the correlation between energy supply and carbon emissions in India within a specific empirical framework. Results revealed that there is a feedback loop between economic growth and emissions, but no long-term causal links. The results also show that changes in the supply of energy happen before changes in carbon emissions and that changes in the economy affect how the supply of energy changes over the same period. Gu et al. (2022) employed a more comprehensive panel framework, emphasizing cross-sectional interactions among nations.

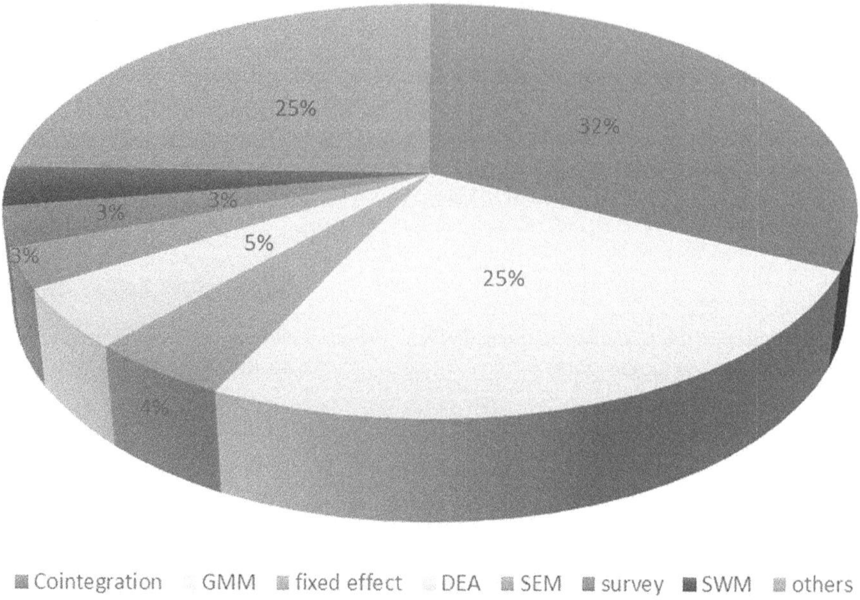

FIGURE 5.3 The distribution of papers by methodology.

Their results demonstrate the effect of the Logistics Performance Index (LPI) on carbon emissions.

Islam et al. (2021) examined the impact of carbon emissions, precipitation, inflation, and unemployment on economic growth in Saudi Arabia, employing the autoregressive distributed lag (ARDL) cointegration methodology to investigate long-term relationships among these variables from 1990 to 2019. The results indicate that Saudi Arabia can attain greater economic growth without adequately tackling unemployment, as variations in carbon emissions adversely affect economic growth.

Khan et al. (2018) investigate the nexus between green logistics services, energy demand, sustainable economic development, and environmental performance using the ordinary least squares (OLS) statistical method. Their results indicate a negative relationship between carbon emissions and green logistics, while green logistics indicators show a strong positive correlation with foreign direct investment (FDI) flows, trade openness, and green energy sources. Their findings demonstrate a favorable long-term correlation between renewable energy and green logistics, whereas fossil fuels exhibit a detrimental relationship with green logistics operations. Menyah and Wolde-Rufael (2010) used the cointegration method and a modified Granger causality test to investigate the long-term relationship between energy use, pollution emissions, and economic growth. They found a positive and statistically significant relationship between economic growth and pollutant emissions, with both short- and long-term links between these variables. Mujtaba and Shahzad (2021) performed an analysis of economic growth, environmental pollutants, and public health. Raggad

(2018) used an ARDL approach with structural breaks to investigate the long-term relationship between economic growth, urbanization, energy use, and CO_2 emissions in Saudi Arabia.

Suki et al. (2021) emphasized the significance of the LPI in reducing carbon emissions and fostering economic growth through various fundamental tests, including cross-sectional dependence, cointegration, unit root tests, and cross-sectionally augmented autoregressive distributed lag (CS-ARDL) regression. Jia et al. (2022) investigated the economic factors of national carbon emissions using robust OLS and Markov switching regression. The results show that better logistics operations help lower carbon emissions. Other important factors include renewable energy, sustainable fuel imports, the cost of starting a business, and information disclosure. Amirnejad et al. (2021) confirmed the relationship between per capita gross domestic product (GDP) as an income indicator and forest energy depletion, validating the environmental Kuznets curve (EKC) hypothesis in both U-shaped and inverted U-shaped forms.

Li et al. (2021) also examined how green logistics performance affects economic growth using two-stage least squares (2SLS) and generalized method of moments (GMM) estimators. Results revealed that green logistics performance improves economic growth in MENA, Europe, and BRI regions. Karaduman et al. (2020) focused on analyzing the impact of logistics performance on carbon (CO_2) emissions by using a fixed-effects model to establish the causal relationship between logistics performance and carbon dioxide emissions in the Balkan countries. Their findings reveal a positive and significant correlation between carbon dioxide emissions and logistics performance.

Mariano et al. (2017) assessed the efficiency of the relationship between transportation logistics performance to develop a Low Carbon Logistics Performance Index (LCLPI). Lu et al. (2019) evaluated the performance of green transportation using the Data Envelopment Analysis (DEA). Their results indicate that emissions of four pollutants initially rise and subsequently decline as economic growth occurs in both countries. Prajogo et al. (2012) conducted an empirical study exploring the relationship between a company's operational performance and its supplier management practices, utilizing structural equation modeling (SEM). The findings suggest that various metrics of process performance are affected by different supplier management practices. The results of Khan (2020) indicate a negative relationship between environmental performance and public healthcare spending.

5.3.5 Logistics Policy Development

Policy development is influenced by interactions within the transport sector. Numerous studies have examined the evolution of policies related to logistics operations, such as congestion in remote areas concerning vertical integration and port activities (De Borger & De Bruyne, 2011). The impact of port decisions on regional economic development and transport and logistics systems, and their influence on investment performance, make port governance a compelling topic for academic inquiry.

Numerous studies focusing on the effect of policies on economic and logistics performance provide valuable insights for policymakers (Lean et al., 2014). Ponti et al. (2013) reviewed the transfer policies and political procedures of the European Union, focusing on regulation, competition, and infrastructure, while also addressing the need for further development of environmental and safety principles. The Global Competitiveness Index positively influences LPI rankings, as corrective measures in this area aim to demonstrate. Any improvement in LPI standards contributes to enhanced competitive strength. Despite this, there remains a limited understanding of how to formulate national policies, logistics, and business strategies based on the results of the LPI, which is regarded as a critical criterion for achieving these objectives.

5.4 CONCLUSION AND POLICY IMPLICATIONS

The existing information about the correlation between logistics performance and economic and environmental outcomes has been fragmented, with the literature revealing regional and methodological limitations. This review closes the gap by systematically aggregating results from 65 high-quality research published between 2010 and 2023, providing the most comprehensive evidence base currently accessible. The LPI shows a strong link to economic results, which confirms its value for policy formulation and research initiatives. When companies construct roads, ports, or warehouses, environmental damage usually rises. By employing green logistics such as optimizing processes, transitioning to more sustainable transport modes, and integrating advanced technology, they safeguard the environment while simultaneously achieving economic benefits. This discovery refutes the prevalent notion that economic advancement necessarily harms the environment. Logistics initiatives that fulfill both economic and environmental objectives occur more frequently than previously assumed, with over one-third of the analyzed studies documenting such instances. Fourth, most of the research originates from Asia and is predominantly quantitative; there is limited understanding of qualitative factors, institutional contexts, or conditions in other regions globally. The findings carry important theoretical implications. They reinforce the EKC hypothesis within the logistics domain. The environmental degradation initially rises with logistics development, but later declines as cleaner technologies are adopted. Through technology transfer, focused policy design, and adoption of advanced countries' best practices, emerging nations can separate logistics growth from environmental degradation, challenging the pollution haven concept. These findings refine logistics–economy–environment nexus theories and emphasize technical and institutional enablers. Geographic coverage is another significant constraint. Many studies derived from Asian contexts, hence constraining the applicability of results to other countries.

This study generates several important policy implications for achieving sustainable logistics development. First, policymakers should focus on the investment in green logistics technologies, prioritizing modal shift, energy efficiency, and renewable energy integration. The research implies that such investments can generate economic and environmental returns. Second, developing nations should learn from developed nations and adopt cleaner technology and methods from the start, rather than "develop first, clean up later". This involves international cooperation, technological transfer, and funding. Fourth, regulations should encourage sustainable logistics without stifling innovation. Carbon pricing and green procurement strategies are promising.

Given the absence of a quantitative analysis in this study, we propose the adoption of meta-analysis as a future research direction. This method offers the advantage of enabling the generalization of results due to the diversity of populations and contexts examined. Moreover, by integrating data from multiple studies, meta-analysis makes it possible to identify effects that may not appear statistically significant in individual studies, thereby strengthening the statistical power of empirical validation.

REFERENCES

Aichele, R., & Felbermayr, G. (2012). Kyoto and the carbon footprint of nations. *Journal of Environmental Economics and Management*, *63*(3), 336–354.

Akram, M. W., Hafeez, M., Yang, S., Sethi, N., Mahar, S., & Salahodjaev, R. (2023). Asian logistics industry efficiency under low carbon environment: Policy implications for sustainable development. *Environmental Science and Pollution Research*, *30*(21), 59793–59801.

Ali, E. B., Shayanmehr, S., Radmehr, R., Amfo, B., Awuni, J. A., Gyamfi, B. A., & Agbozo, E. (2023). Exploring the impact of economic growth on environmental pollution in South American countries: How does renewable energy and globalization matter?. *Environmental Science and Pollution Research*, *30*(6), 15505–15522.

Ali, S., Ying, L., Anjum, R., Nazir, A., Shalmani, A., Shah, T., & Shah, F. (2021). Analysis on the nexus of CO_2 emissions, energy use, net domestic credit, and GDP in Pakistan: An ARDL bound testing analysis. *Environmental Science and Pollution Research*, *28*(4), 4594–4614.

Alvarado, R., & Toledo, E. (2017). Environmental degradation and economic growth: Evidence for a developing country. *Environment, Development and Sustainability*, *19*(4), 1205–1218.

Amirnejad, H., Mehrjo, A., & Yuzbashkandi, S. S. (2021). Economic growth and air quality influences on energy sources depletion, forest sources and health in MENA. *Environmental Challenges*, *2*, 100011.

An, H., Razzaq, A., Nawaz, A., Noman, S. M., & Khan, S. A. R. (2021). Nexus between green logistic operations and triple bottom line: Evidence from infrastructure-led Chinese outward foreign direct investment in Belt and Road host countries. *Environmental Science and Pollution Research*, *28*(37), 51022–51045.

Aydin, C., Aydin, H., & Altinok, H. (2023). Does the level of energy intensity matter in the effect of logistic performance on the environmental pollution of OBOR countries? Evidence from PSTR analysis. *Journal of Environmental Planning and Management*, *66*(7), 1494–1512.

Bansal, S., Sharma, G. D., Rahman, M. M., Yadav, A., & Garg, I. (2021). Nexus between environmental, social and economic development in South Asia: Evidence from econometric models. *Heliyon*, *7*(1), e05965.

Barrios, S., Bertinelli, L., & Strobl, E. (2010). Trends in rainfall and economic growth in Africa: A neglected cause of the African growth tragedy. *The Review of Economics and Statistics*, *92*(2), 350–366.

Bello, A. K., & Abimbola, O. M. (2010). Does the level of economic growth influence environmental quality in Nigeria: A test of environmental Kuznets curve (EKC) hypothesis. *Pakistan Journal of Social Sciences*, *7*(4), 325–329.

Bölük, G., & Mert, M. (2015). The renewable energy, growth and environmental Kuznets curve in Turkey: An ARDL approach. *Renewable and Sustainable Energy Reviews*, *52*, 587–595.

Borhan, H., Ahmed, E. M., & Hitam, M. (2012). The impact of CO_2 on economic growth in ASEAN 8. *Procedia-Social and Behavioral Sciences*, *35*, 389–397.

Büyüközkan, G., & Çifçi, G. (2011). A novel fuzzy multi-criteria decision framework for sustainable supplier selection with incomplete information. *Computers in Industry*, *62*(2), 164–174.

Chu, Z. (2012). Logistics and economic growth: A panel data approach. *The Annals of Regional Science*, *49*(1), 87–102.

Coto-Millán, P., Agüeros, M., Casares-Hontañón, P., & Pesquera, M. Á. (2013). Impact of logistics performance on world economic growth (2007–2012). *World Review of Intermodal Transportation Research*, *4*(4), 300–310.

De Borger, B., & De Bruyne, D. (2011). Port activities, hinterland congestion, and optimal government policies the role of vertical integration in logistic operations. *Journal of Transport Economics and Policy (JTEP)*, *45*(2), 247–275.

Dell, M., Jones, B. F., & Olken, B. A. (2012). Temperature shocks and economic growth: Evidence from the last half century. *American Economic Journal: Macroeconomics*, *4*(3), 66–95.

Ding, S., Ward, H., & Tukker, A. (2023). How Internet of Things can influence the sustainability performance of logistics industries – A Chinese case study. *Cleaner Logistics and Supply Chain*, *6*, 100094.

Fan, M., Wu, Z., Qalati, S. A., He, D., & Hussain, R. Y. (2022). Impact of green logistics performance on China's export trade to regional comprehensive economic partnership countries. *Frontiers in Environmental Science*, *10*, 879590.

Ghosh, S. (2010). Examining carbon emissions economic growth nexus for India: A multivariate cointegration approach. *Energy Policy*, *38*(6), 3008–3014.

Göçer, A., Özpeynirci, Ö., & Semiz, M. (2022). Logistics performance index-driven policy development: An application to Turkey. *Transport Policy*, *124*, 20–32.

Godil, D. I., Yu, Z., Sharif, A., Usman, R., & Khan, S. A. R. (2021). Investigate the role of technology innovation and renewable energy in reducing transport sector CO_2 emission in China: A path toward sustainable development. *Sustainable Development*, *29*(4), 694–707.

Gu, Z., Malik, H. A., Chupradit, S., Albasher, G., Borisov, V., & Murtaza, N. (2022). Green supply chain management with sustainable economic growth by CS-ARDL technique: Perspective to blockchain technology. *Frontiers in Public Health*, *9*, 818614.

Hao, Y., Gao, S., Guo, Y., Gai, Z., & Wu, H. (2021). Measuring the nexus between economic development and environmental quality based on environmental Kuznets curve: A comparative study between China and Germany for the period of 2000–2017. *Environment, Development and Sustainability*, *23*(11), 16848–16873.

Hayaloğlu, P. (2015). The impact of developments in the logistics sector on economic growth: The case of OECD countries. *International Journal of Economics and Financial Issues*, *5*(2), 523–530.

He, J., & Richard, P. (2010). Environmental Kuznets curve for CO_2 in Canada. *Ecological Economics*, *69*(5), 1083–1093.

He, Z., Chen, P., Liu, H., & Guo, Z. (2017). Performance measurement system and strategies for developing low-carbon logistics: A case study in China. *Journal of Cleaner Production*, *156*, 395–405.

Islam, M. M., Alharthi, M., & Murad, M. W. (2021). The effects of carbon emissions, rainfall, temperature, inflation, population, and unemployment on economic growth in Saudi Arabia: An ARDL investigation. *PLoS One*, *16*(4), e0248743.

Jia, X., Dong, F., & Sun, B. (2022). Economic determinants of national carbon emissions: Perspectives from 119 countries. *Science of the Total Environment*, 835, 155553.

Jiang, X., Ma, J., Zhu, H., Guo, X., & Huang, Z. (2020). Evaluating the carbon emissions efficiency of the logistics industry based on a super-SBM model and the Malmquist Index from a strong transportation strategy perspective in China. *International Journal of Environmental Research and Public Health, 17*(22), 8459.

Jomthanachai, S., Wong, W. P., & Khaw, K. W. (2022). An application of machine learning regression to feature selection: A study of logistics performance and economic attribute. *Neural Computing and Applications, 34*(18), 15781–15805.

Karaduman, H. A., Karaman-Akgül, A., Çağlar, M., & Akbaş, H. E. (2020). The relationship between logistics performance and carbon emissions: An empirical investigation on Balkan countries. *International Journal of Climate Change Strategies and Management, 12*(4), 449–461.

Khan, M. (2020). CO_2 emissions and sustainable economic development: New evidence on the role of human capital. *Sustainable Development, 28*(5), 1279–1288.

Khan, S. A. R. (2019). The nexus between carbon emissions, poverty, economic growth, and logistics operations-empirical evidence from Southeast Asian countries. *Environmental Science and Pollution Research, 26*(13), 13210–13220.

Khan, S. A. R., Dong, Q. L., & Yu, Z. (2016). Study of logistics and manufacturing industry integration from the perspective of Pakistan. *International Journal of Engineering Research in Africa, 24*, 172–180.

Khan, S. A. R., Jian, C., Zhang, Y., Golpîra, H., Kumar, A., & Sharif, A. (2019). Environmental, social and economic growth indicators spur logistics performance: From the perspective of South Asian Association for Regional Cooperation countries. *Journal of Cleaner Production, 214*, 1011–1023.

Khan, S. A. R., & Qianli, D. (2017). Does national scale economic and environmental indicators spur logistics performance? Evidence from UK. *Environmental Science and Pollution Research, 24*(34), 26692–26705.

Khan, S. A. R., Qianli, D., SongBo, W., Zaman, K., & Zhang, Y. (2017). Environmental logistics performance indicators affecting per capita income and sectoral growth: Evidence from a panel of selected global ranked logistics countries. *Environmental Science and Pollution Research, 24*(2), 1518–1531.

Khan, S. A. R., Zhang, Y., Anees, M., Golpîra, H., Lahmar, A., & Qianli, D. (2018). Green supply chain management, economic growth and environment: A GMM based evidence. *Journal of Cleaner Production, 185*, 588–599.

Khan, S. A. R., Zia-ul-haq, H. M., Umar, M., & Yu, Z. (2021). Digital technology and circular economy practices: An strategy to improve organizational performance. *Business Strategy & Development, 4*(4), 482–490.

Khan, Z., Badeeb, R. A., & Nawaz, K. (2022). Natural resources and economic performance: Evaluating the role of political risk and renewable energy consumption. *Resources Policy, 78*, 102890.

Kim, I., & Min, H. (2011). Measuring supply chain efficiency from a green perspective. *Management Research Review, 34*(11), 1169–1189.

Kusi-Sarpong, S., Gupta, H., & Sarkis, J. (2019). A supply chain sustainability innovation framework and evaluation methodology. *International Journal of Production Research, 57*(7), 1990–2008.

Larson, P. D. (2021). Relationships between logistics performance and aspects of sustainability: A cross-country analysis. *Sustainability, 13*(2), 623.

Lean, H. H., Huang, W., & Hong, J. (2014). Logistics and economic development: Experience from China. *Transport Policy, 32*, 96–104.

Li, X., Sohail, S., Majeed, M. T., & Ahmad, W. (2021). Green logistics, economic growth, and environmental quality: Evidence from one belt and road initiative economies. *Environmental Science and Pollution Research, 28*(24), 30664–30674.

Lin, X., Zhao, Y., Ahmad, M., Ahmed, Z., Rjoub, H., & Adebayo, T. S. (2021). Linking innovative human capital, economic growth, and CO_2 emissions: An empirical study based on Chinese provincial panel data. *International Journal of Environmental Research and Public Health, 18*(16), 8503.

Liu, J., Yuan, C., Hafeez, M., & Yuan, Q. (2018). The relationship between environment and logistics performance: Evidence from Asian countries. *Journal of Cleaner Production, 204*, 282–291.

Liu, L., Wu, H., Hafeez, M., Abubaker Albaity, M. S., & Ullah, S. (2023). Carbon neutrality through supply chain performance: Does green innovation matter in Asia? *Economic Research-Ekonomska istraživanja, 36*(3).

Lu, M., Xie, R., Chen, P., Zou, Y., & Tang, J. (2019). Green transportation and logistics performance: An improved composite index. *Sustainability, 11*(10), 2976.

Mariano, E. B., Gobbo Jr, J. A., de Castro Camioto, F., & do Nascimento Rebelatto, D. A. (2017). CO_2 emissions and logistics performance: A composite index proposal. *Journal of Cleaner Production, 163*, 166–178.

Menyah, K., & Wolde-Rufael, Y. (2010). Energy consumption, pollutant emissions and economic growth in South Africa. *Energy Economics, 32*(6), 1374–1382.

Mohsin, A. K. M., Tushar, H., Hossain, S. F. A., Chisty, K. K. S., Iqbal, M. M., Kamruzzaman, M., & Rahman, S. (2022a). Retracted: Green logistics and environment, economic growth in the context of the belt and road initiative. *Heliyon, 8*(6), e09641.

Mohsin, M., Taghizadeh-Hesary, F., Iqbal, N., & Saydaliev, H. B. (2022b). The role of technological progress and renewable energy deployment in green economic growth. *Renewable Energy, 190*, 777–787.

Moreira, O. J., & Rodrigues, M. C. M. (2023). Sourcing third party logistics service providers based on environmental, social and corporate governance: A case study. *Discover Sustainability, 4*(1), 36.

Mujtaba, G., & Shahzad, S. J. H. (2021). Air pollutants, economic growth and public health: Implications for sustainable development in OECD countries. *Environmental Science and Pollution Research, 28*(10), 12686–12698.

Odusola, A., & Abidoye, B. (2015). *Effects of temperature and rainfall shocks on economic growth in Africa. 29th Triennial Conference of the International Association of Agricultural Economists (IAAE)*, Milan, Italy.

Ouni, M., Ben Abdallah, K., & Ouni, F. (2023). The nexus between indicators for sustainable transportation: A systematic literature review. *Environmental Science and Pollution Research, 30*(42), 95272–95295.

Peng, B., Sheng, X., & Wei, G. (2020). Does environmental protection promote economic development? From the perspective of coupling coordination between environmental protection and economic development. *Environmental Science and Pollution Research, 27*(31), 39135–39148.

Ponti, M., Boitani, A., & Ramella, F. (2013). The European transport policy: Its main issues. *Case Studies on Transport Policy, 1*(1–2), 53–62.

Prajogo, D., Chowdhury, M., Yeung, A. C., & Cheng, T. C. E. (2012). The relationship between supplier management and firm's operational performance: A multi-dimensional perspective. *International Journal of Production Economics, 136*(1), 123–130.

Raggad, B. (2018). Carbon dioxide emissions, economic growth, energy use, and urbanization in Saudi Arabia: Evidence from the ARDL approach and impulse saturation break tests. *Environmental Science and Pollution Research, 25*(15), 14882–14898.

Rashidi, K., & Cullinane, K. (2019). Evaluating the sustainability of national logistics performance using data envelopment analysis. *Transport Policy, 74*, 35–46.

Saidi, S., Mani, V., Mefteh, H., Shahbaz, M., & Akhtar, P. (2020). Dynamic linkages between transport, logistics, foreign direct Investment, and economic growth: Empirical evidence from developing countries. *Transportation Research Part A: Policy and Practice, 141*, 277–293.

Suki, N. M., Suki, N. M., Sharif, A., & Afshan, S. (2021). The role of logistics performance for sustainable development in top Asian countries: Evidence from advance panel estimations. *Sustainable Development*, *29*(4), 595–606.

Wan, B., Wan, W., Hanif, N., & Ahmed, Z. (2022). Logistics performance and environmental sustainability: Do green innovation, renewable energy, and economic globalization matter?. *Frontiers in Environmental Science*, *10*, 996341.

Wang, W., & Wu, Q. (2023). Research on coordinated development of Shenzhen port logistics and hinterland economy. *Sustainability*, *15*(5), 4083.

Yu, W., Chavez, R., Jacobs, M. A., & Feng, M. (2018). Data-driven supply chain capabilities and performance: A resource-based view. *Transportation Research Part E: Logistics and Transportation Review*, *114*, 371–385.

Yunlin, C. (2023). Awareness of green logistics technology, certification, and standards by logistics practitioners at Chinese e-commerce company, Jing Dong. *The Asian Journal of Shipping and Logistics*, *39*(4), 37–46.

Zaman, K., & Shamsuddin, S. (2017). Green logistics and national scale economic indicators: Evidence from a panel of selected European countries. *Journal of Cleaner Production*, *143*, 51–63.

Section III

Technology and Digital Transformation

6 A Theoretical Framework for Integrating Social Media Marketing into Supply Chain Management

Perspective from RBV, DCV, and TCE

Lawal Abideen Ayokunmi, Noor Aslinda Abu Seman, and Umi Kartini Rashid

6.1 INTRODUCTION

In the contemporary digital economy, social media marketing (SMM) has evolved from a promotional tool to a core mechanism for organizational communication and coordination in both B2B and B2C contexts (Saddhono et al., 2020). For resource-constrained SMEs, SMM provides a cost-effective platform to enhance communication, optimize logistics, and engage stakeholders. While traditional supply chains rely on conventional communication and transaction models, digital platforms are increasingly embedded in supply chain strategies to streamline processes, mitigate risks, and improve competitiveness (Abou Kamar et al., 2023). However, SCM research remains largely operational, with limited integration of SMM into theoretical frameworks, leaving a gap in understanding its contribution to efficiency, resilience, and cost-effectiveness. Although the Resource-Based View (RBV), Dynamic Capabilities View (DCV), and Transaction Cost Economics (TCE) are widely applied in management research, their combined application to SCM and SMM is rare (Dwivedi et al., 2022). This fragmentation underscores the need for integrative studies positioning SMM as a strategic resource (RBV), a driver of dynamic capabilities (DCV), and a mechanism to reduce transaction costs (TCE). Thus, this

DOI: 10.1201/9781003770008-9

study aims to develop a conceptual framework bridging these perspectives to explain SMM's role in SCM, guided by three central research questions:

1. How can SMM be conceptualized as a strategic resource in SCM (RBV)?
2. How does SMM enhance resilience and agility in SCM (DCV)?
3. How does SMM reduce transaction costs across supply chains (TCE)?

The RBV provides a foundation for understanding how firms gain competitive advantage through valuable, rare, inimitable, and non-substitutable (VRIN) resources (Porter, 1980, 1985). When SMM is applied to SCM, it represents a strategic resource that enhances information flow, strengthens supplier relationships, and increases responsiveness to customer needs (Filho & Moori, 2020; Cao et al., 2021; Dwivedi et al., 2022). Building on this, the DCV stresses the need for firms to remain agile in turbulent environments (Teece et al., 1997; Ramos et al., 2023). By facilitating coordination and rapid responses to disruptions, SMM enables supply chains to adapt to uncertainty and build resilience, particularly during crises such as the COVID-19 pandemic (Mas Suariedewi et al., 2022; Yusriadi et al., 2022). Meanwhile, TCE emphasizes efficiency by minimizing negotiation, monitoring, and coordination costs (Lee & Zhong, 2020). Through direct and transparent interactions, SMM reduces transaction costs, fosters trust, and strengthens collaboration among partners (Filho & Moori, 2020). However, these frameworks also reveal tensions: RBV overlooks cost structures, DCV underplays the rarity of resources, and TCE neglects the strategic value of digital tools. Therefore, synthesizing them provides a more holistic explanation of SMM's role in SCM (Ramos et al., 2023). This study contributes theoretically by integrating RBV, DCV, and TCE in SMM–SCM research, linking strategic resources, adaptability, and efficiency. Practically, it guides SMEs and supply chain professionals in leveraging SMM to build resilient, sustainable supply chains and informs policymakers supporting digital transformation post-pandemic. This paper proceeds as follows: Section 6.2 reviews literature on SMM, SCM, and RBV, DCV, and TCE; Section 6.3 details methodology; Section 6.4 presents findings; Section 6.5 discusses theoretical and practical implications; and Section 6.6 concludes with key contributions and future research directions.

6.2 LITERATURE REVIEW

An Overview of Social Media Marketing in Supply Chain Management Integrating SMM into supply chain management (SCM) has transformed traditional business operations, facilitating enhanced communication, transparency, and efficiency (Saddhono et al., 2020; Yusriadi et al., 2022). Businesses are increasingly using social media platforms such as Facebook, LinkedIn, Twitter, and Instagram to streamline supply chain processes, improve supplier relationships, and enhance customer engagement (Saddhono et al., 2020). As digital technologies continue to evolve, understanding the intersection between SMM and SCM becomes critical for organizations seeking to optimize logistics, strengthen partnerships, and adapt to dynamic market conditions (Uddin et al., 2023). This literature review explores existing research on the role of SMM in SCM, identifying key themes, challenges, and opportunities within this growing field.

6.2.1 The Role of Social Media Marketing in Supply Chain Management

Social media enhances communication among supply chain stakeholders, overcoming delays associated with traditional channels like emails and meetings (Ojubanire & Popoola, 2022). Real-time interactions enable suppliers, manufacturers, and distributors to coordinate efficiently, track shipments, monitor production schedules, and manage inventory, improving responsiveness (Saddhono et al., 2020; Yusriadi et al., 2022; Uddin et al., 2023). Collaboration is strengthened through transparent engagement with suppliers, vendors, and customers, while platforms like LinkedIn, Twitter, and Facebook facilitate networking and information sharing (Ojong-Ejoh et al., 2021; Xu & Zhao, 2022). Social media also supports market intelligence and demand forecasting by analyzing discussions, trends, and feedback, helping firms adjust strategies, minimize overproduction, and optimize inventory (Alkhateri et al., 2021; Nasrollahi, 2018; Basri & Siam, 2019; Cao et al., 2021). During disruptions from geopolitical, economic, or natural events, social media enables agile crisis communication and coordination of alternative logistics solutions (Thongrawd et al., 2020; Singh et al., 2021).

Furthermore, SMM fosters proactive risk management, enhancing resilience and adaptability in unpredictable markets (Dwivedi et al., 2022). Transparent communication of ethical sourcing, sustainability, and corporate responsibility initiatives improves brand reputation and consumer trust (Basri & Siam, 2019; Mas Suariedewi et al., 2022). Direct engagement through social media addresses concerns, gathers feedback, and strengthens long-term customer relationships and brand advocacy (Wijaya et al., 2021; Singh et al., 2021).

6.2.2 Challenges of Integrating Social Media into Supply Chain Management

Although social media enhances SCM through improved communication, market intelligence, and real-time decision-making, its integration poses several challenges (Orji et al., 2020). Data overload is a primary concern, as platforms generate massive volumes of information that can overwhelm decision-making without proper filtering (Ninib et al., 2020). Firms must employ advanced data management systems, AI, and machine learning to extract meaningful insights and optimize supply chain operations (Martínez-López et al., 2022). Cybersecurity and privacy risks are also critical, as sensitive supply chain information is exposed to potential cyber threats, including data breaches and intellectual property theft (Cheung et al., 2020; Chatterjee et al., 2021). Organizations must implement strong cybersecurity protocols, encrypted communications, and regular risk assessments to safeguard operations (Swallehe & Haule, 2021). Additionally, the reliability of social media data is a challenge, given the prevalence of misinformation and biased user-generated content (Cheung et al., 2020; Wali et al., 2020). Therefore, rigorous data validation, cross-referencing sources, and integrating human expertise are essential to ensure informed decisions (Cheung et al., 2020). Addressing these challenges allows businesses to leverage social media effectively for efficient and resilient supply chains (Martínez-López et al., 2022).

6.2.3 Scope and Limitations of Reviewed Studies

This literature review synthesizes studies published between 2017 and 2025 in peer-reviewed journals indexed in Scopus, PubMed, ScienceDirect, and Web of Science. The focus is placed on small and medium-sized enterprises (SMEs), which represent the backbone of most economies, yet remain disproportionately vulnerable during crises such as the COVID-19 pandemic. By consolidating recent scholarly contributions, the review highlights evolving perspectives on resilience, innovation, and digital adoption as critical drivers of SME survival and growth. While this provides up-to-date insights, the review is constrained by the limited empirical validation of certain conceptual frameworks across diverse contexts and by a heavy reliance on secondary sources rather than original field data. Moreover, the findings predominantly reflect trends in emerging literature, which may not always translate seamlessly into practice. Future research would benefit from comparative analyses across industries and regions to strengthen external validity. Equally, there is a pressing need for longitudinal studies that capture dynamic changes over time, offering a more holistic understanding of SME adaptation and performance.

6.2.4 Theoretical Foundations in Supply Chain Management

In the digital age, businesses are increasingly turning to social media as a strategic tool to enhance SCM, adapt to dynamic market conditions, and reduce operational costs (Shibin et al., 2020). The integration of social media into SCM is best understood through the lens of RBV, DCV, and TCE. These theoretical frameworks offer valuable insights into how businesses can leverage digital platforms to strengthen supply chain resilience, optimize decision-making, and gain competitive advantages (Uddin et al., 2023). This literature review explores the application of RBV, DCV, and TCE theories in the context of social media-driven SCM, highlighting their relevance in improving supply chain efficiency, fostering adaptability, and minimizing transaction costs.

6.2.4.1 Transaction Cost Economics Theory

The concept of transaction costs was initially introduced by Coase (1937) and was later formalized by Williamson (1975, 1979), who defined TCE as the study of minimizing total transaction costs through optimal governance structures, either hierarchical or market-based (Williamson, 1979; Inge et al., 2006; Lee & Zhong, 2020). Hierarchical governance manages transactions internally, while market governance relies on supply-and-demand mechanisms. The choice between internal and external transactions depends on complexities and costs, including contractual hazards under uncertainty such as bounded rationality, opportunism, small-number bargaining, and information asymmetry (Grover & Malhotra, 2003; Williamson, 2010; Garfamy, 2012). In addition, transactional costs encompass negotiation, control, communication, and contract writing (Neves et al., 2014). Williamson (1979) identified asset specificity, uncertainty, and transaction frequency as key factors that increase costs

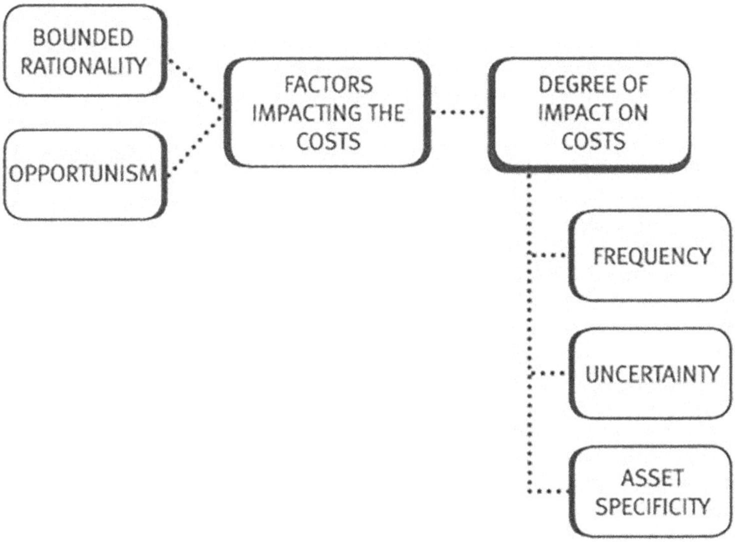

FIGURE 6.1 Transaction cost economics.

and favor hierarchical governance (Inge et al., 2006). Asset specificity determines the redeployability of resources, uncertainty arises from unpredictable changes requiring renegotiation, and high-frequency transactions are often more efficiently conducted internally (Williamson, 1979). In this study, TCE is seen as vital for understanding organizational governance structures, whether market or hierarchical. The decision to perform a transaction internally or externally is determined by the related problems and costs, with market arrangements chosen for low-cost transactions and hierarchical systems preferred for high-cost transactions. Figure 6.1 illustrates the TCE theory.

6.2.4.2 Resource-Based View Theory

For a long time, it was widely believed that industrial organizations played a crucial role in identifying competitive strategies that could enhance business performance (Porter, 1980, 1985). Porter (1985), argued that competitive advantage depends on both industry attractiveness and a firm's position within the industry, while the ability to deliver superior value to customers is critical for sustaining that advantage. However, industry conditions alone do not fully explain performance differences (Ellmarker, 2017). Grant (1991) showed that variations are often greater within industries, highlighting the importance of internal resources over external conditions (Madhani, 2010). Effectively managed resources can create entry barriers, strengthen bargaining power, limit substitutes, and influence market competition (Filho & Moori, 2020). Furthermore, Barney (1991) defined resources as the internal capabilities within a firm, while Wernerfelt (1984) proposed that resources, both tangible and intangible, represent a company's strengths and weaknesses, drawing from Andrews'

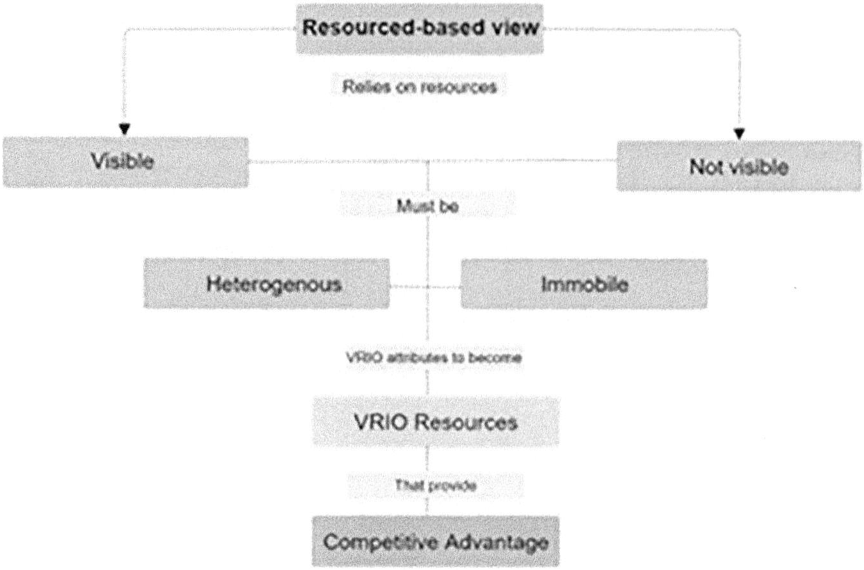

FIGURE 6.2 RBV.

(1971) SWOT paradigm. Strategic application of resources is required to generate value, as uncoordinated resources may not yield financial gains (Gavronski et al., 2006; Filho & Moori, 2020). Returns depend on resource utilization strategies coordinated through organizational routines and operational capabilities, which evolve via learning and collaboration (Grant, 1996; Teece et al., 1997; Neves et al., 2014). The RBV theory, illustrated in Figure 6.2, underscores the importance of leveraging internal assets to gain a sustained competitive advantage.

In recent years, SCM has become a critical element of strategic management, enhancing competitiveness in the global marketplace (Filho & Moori, 2020). SCM offers new analytical perspectives and organizational structures that improve operational agility and efficiency through information systems (Duong & Nguyen, 2018). The RBV framework supports agility, adaptability, and alignment in SCM, emphasizing key resources such as heterogeneity, allocation, utilization, and imitability to develop distinctive capabilities that boost performance (Dubey et al., 2018; Shibin et al., 2020; Thongrawd et al., 2020). Collaborative relationships across the supply chain are increasingly prioritized to enhance overall performance (Xu & Zhao, 2022). The RBV framework emphasizes the utilization of both tangible and intangible assets, including human and non-human resources, to create added value in products and services (Wernerfelt, 1984; Barney, 1991). Integration and collaboration in internal and external supply chain activities further optimize performance (Prajogo & Olhager, 2012; Birasnav & Bienstock, 2019; Munir et al., 2020). Strategic partnerships require investments in shared resources, information exchange, and equitable distribution of responsibilities (Xu & Zhao, 2022). Despite progress, research gaps remain in understanding collaboration across supply chain links and its effect on superior performance, highlighting opportunities for enhancing sustainability and relationship management (Shibin et al., 2020).

6.2.4.3 Dynamic Capabilities View Theory

The DCV theory extends the RBV by emphasizing a firm's ability to adapt its assets and processes to changing environments (Wernerfelt, 1984; Barney, 1991; Teece et al., 1997). Dynamic capabilities enable firms to integrate, reconfigure, and renew internal and external skills, enhancing flexibility and competitiveness in volatile markets (Schilke et al., 2018; Teece, 2018; Abou Kamar et al., 2023; Heubeck, 2023). Firms with DCs can respond strategically to uncertainty by acquiring and deploying resources effectively, leveraging real-time market intelligence, and improving performance (Eisenhardt & Martin, 2000; Teece, 2007; Zahoor et al., 2022). SMEs, particularly affected by COVID-19, faced disrupted supply chains and demand fluctuations, requiring rapid reorganization to remain competitive (Ojong-Ejoh et al., 2021; Singh et al., 2021; Ojubanire & Popoola, 2022). Building dynamic capabilities allows firms to adjust SCM practices, enhance marketing strategies, and improve communication and transformation (Teece et al., 1997; Venciūtė, 2018; Ashiru et al., 2023; Ramos et al., 2023; Uddin et al., 2023).

In addition, SMM provides access to textual data and interactive technologies, supporting innovation, capability development, and competitive advantage (Saddhono et al., 2020; Tungande et al., 2020; Fan et al., 2021; Sengupta et al., 2022; Yusriadi et al., 2022). Thus, integrating SMM with SCM enhances SMEs' resilience, competitiveness, and performance, particularly in post-pandemic contexts (Uddin et al., 2023). In other words, SCM should be coupled with SMM capabilities to increase a firm's competitive edge (Saddhono et al., 2020; Tungande et al., 2020). Figure 6.3 illustrates the dynamic capabilities of SMEs' supply chain in the aftermath of the COVID-19 pandemic.

Figures 6.1–6.3 illustrate how TCE, RBV, and the DCV emphasize different units of analysis and levels of change. While TCE initially addressed market factors, RBV shifted focus to strategic management, and DCV emerged to tackle challenges in rapidly changing global environments. Figure 6.4 depicts this theoretical evolution and the progression toward dynamic perspectives.

In a nutshell, SMM principles may help supply chains develop distinctive and valued knowledge. Advocates of these theories claim that organizations must improve their abilities and resources to beat their competitors (Wernerfelt, 1984; Barney, 1991; Teece et al., 1997; Neves et al., 2014). Previous studies have used the TCE, RBV, and DCV lenses to investigate how the internet and online technologies affect the performance of small and medium-sized organizations. Their findings show that applying SMM concepts not only improves supply chain operational and marketing performance but also encourages the creation of unique resources, hence enhancing their competitive position (Kashurnikov et al., 2019; Hartono et al., 2020; Saddhono et al., 2020; Xu et al., 2021; Gao et al., 2023).

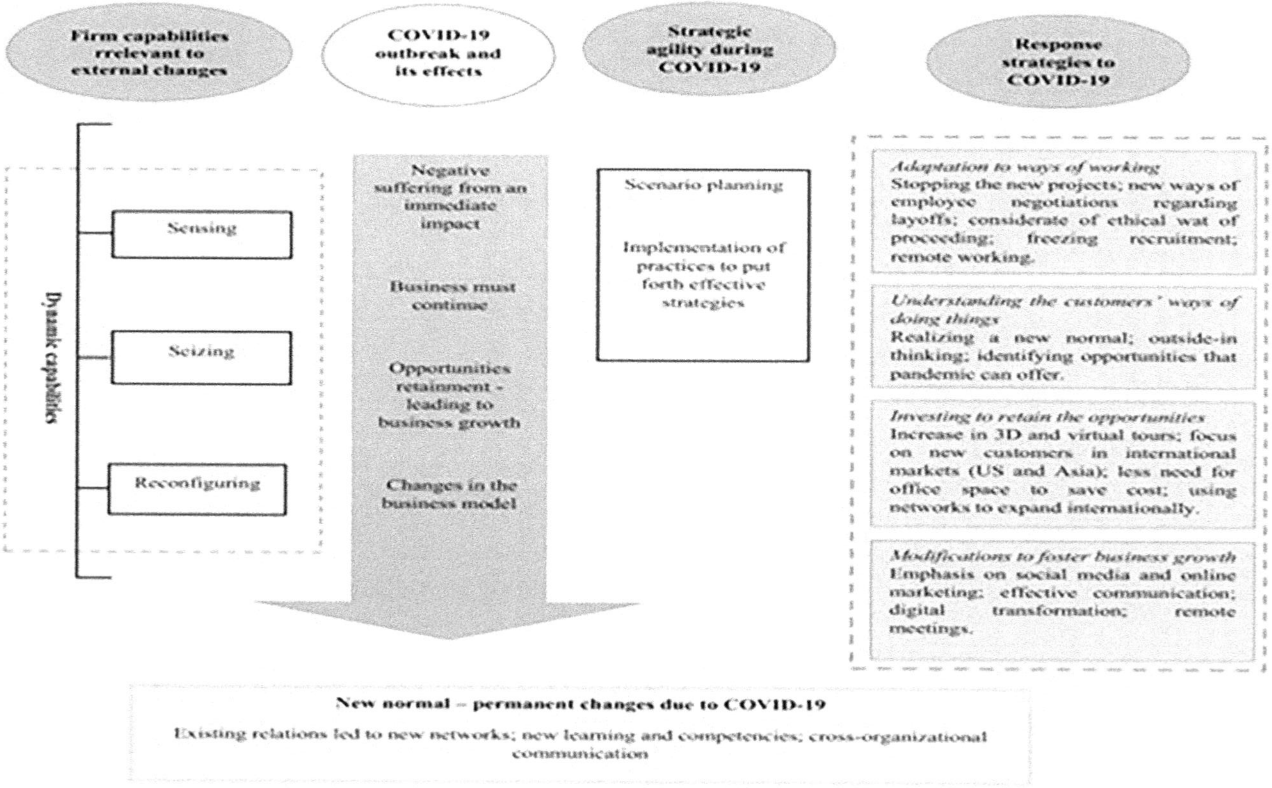

FIGURE 6.3 Dynamic capabilities of SMEs SC.

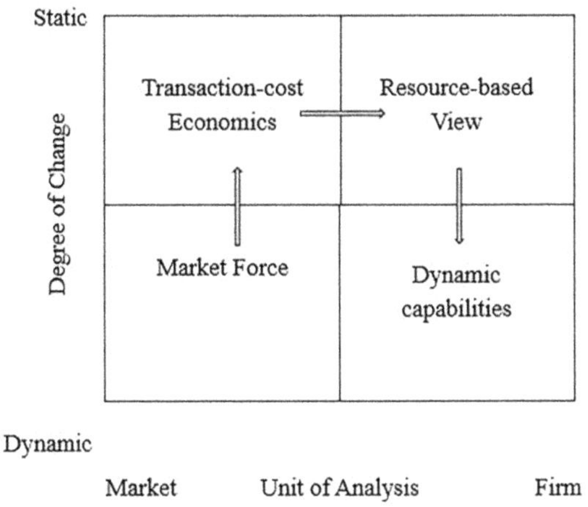

FIGURE 6.4 TCE, RBV, and DCV.

6.3 METHODOLOGY

This study adopts a structured literature review approach to synthesize existing knowledge on the role of SMM in SCM. Unlike a traditional narrative review, a structured approach ensures transparency, replicability, and systematic coverage of relevant studies (Snyder, 2019). The methodology followed the Preferred Reporting Items for Systematic Reviews and Meta-Analyses (PRISMA) guidelines, adapted for conceptual research. Relevant academic publications were identified through searches in Scopus and Web of Science databases, given their reputation for indexing high-quality peer-reviewed journals. The search strategy combined keywords such as: "social media marketing" AND "supply chain management," "digital communication" AND "supply chain resilience," "transaction cost economics" AND "SCM," "resource-based view" OR "dynamic capabilities" AND "social media." The review considered studies published between 2017 and 2025 to capture the most recent insights, especially those reflecting post-pandemic supply chain realities. Furthermore, studies focusing solely on consumer marketing without SCM linkages, as well as non-academic sources such as blogs and news articles, were excluded. The PRISMA flow process ensured that only relevant and high-quality sources contributed to the synthesis. A thematic analysis was employed to classify findings into three theoretical categories: (1) Resource-Based View (RBV): Studies conceptualizing SMM as a strategic resource for SCM. (2) Dynamic Capabilities View (DCV): Studies linking SMM to adaptability, agility, and resilience in supply chains. (3) Transaction Cost Economics (TCE): Studies examining how SMM reduces coordination, monitoring, and negotiation costs. Hence, the analysis followed an iterative coding process, allowing for cross-comparison of themes and the identification of research gaps. The initial query yielded 312 papers: 40 papers from PubMed, 65 articles from the Web of Science database, 93 from ScienceDirect, and 114 from Scopus. The articles published until 2025 were filtered, resulting in three categories. In the first part, 55 duplicate papers were identified, leaving a total of 257. In the second part, after reviewing

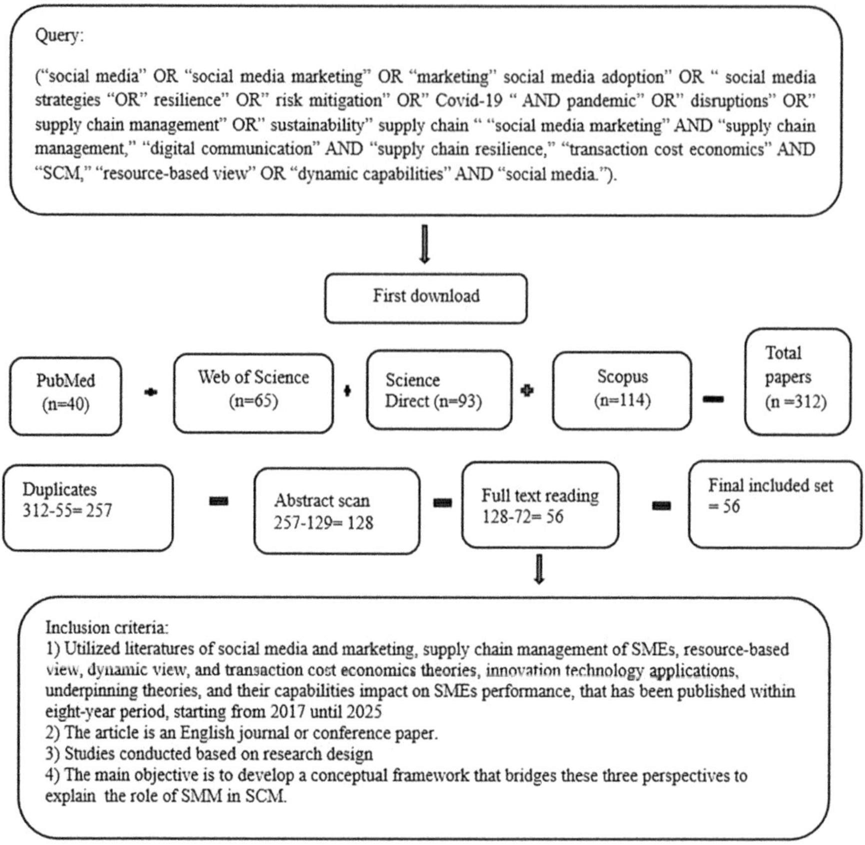

FIGURE 6.5 Flowchart SLR.

titles and abstracts, 129 papers were excluded, reducing the count to 128. The third filtering stage further eliminated 72 papers from the 128, resulting in a final selection of 56 papers, as illustrated in Figure 6.5.

6.4 FINDINGS

The thematic synthesis of the reviewed literature highlights three major insights into the role of SMM in SCM. These findings are organized around the RBV, the DCV, and TCE.

6.4.1 Social Media Marketing as a Strategic Resource (RBV)

From the RBV perspective, SMM functions as a VRIN resource that enhances supply chain competitiveness. The literature demonstrates that firms adopting SMM can improve information flow between suppliers, distributors, and customers (Cao et al.,

2021), and strengthen supplier relationships through consistent and transparent communication (Dwivedi et al., 2022; Xu & Zhao, 2022). They differentiate themselves by leveraging real-time customer insights and data analytics to respond to changing demand (Filho & Moori, 2020). Thus, SMM provides SMEs with a low-cost but strategic tool that enhances visibility, efficiency, and customer engagement across the supply chain.

6.4.2 Social Media Marketing as an Enabler of Resilience and Agility (DCV)

The DCV highlights adaptability as a cornerstone of sustainable supply chains. Findings show that SMM enhances supply chain resilience by enabling firms to coordinate effectively with suppliers and partners during disruptions such as the COVID-19 pandemic (Martínez-López et al., 2022; Uddin et al., 2023). Social media marketing (SMM) enhances supply chain agility by enabling firms to adjust inventory levels and distribution networks in response to demand fluctuations (Saddhono et al., 2020; Yusriadi et al., 2022). It also fosters supply chain resilience through rapid problem-solving and continuity planning supported by digital communicationcommunication (Mas Suariedewi et al., 2022). Thus, SMM equips firms with dynamic capabilities that allow them to reconfigure operations and maintain supply chain continuity under uncertainty.

6.4.3 Social Media Marketing as a Transaction Cost Reducer (TCE)

From the TCE perspective, SMM reduces costs associated with transactions, negotiations, and monitoring by offering direct communication channels between supply chain stakeholders, thereby reducing reliance on intermediaries (Lee & Zhong, 2020). Enhancing trust and transparency reduces the risks of opportunism and information asymmetry (Filho & Moori, 2020). SMM streamlines contract negotiations, shipment tracking, and customer service processes by facilitating instant digital interactions among supply chain partners.(Dwivedi et al., 2022; Xu & Zhao, 2022). These functions are particularly valuable for SMEs that operate with limited financial and managerial resources (Saddhono et al., 2020; Yusriadi et al., 2022).

6.4.4 Integrated Conceptual Insights

Synthesizing these findings reveals that SMM simultaneously operates as a strategic resource that supports competitiveness (RBV), a capability enabler that builds agility and resilience (DCV), and a cost-reduction mechanism that improves efficiency and trust (TCE). While each theoretical lens provides valuable insights, integrating RBV, DCV, and TCE offers a more holistic understanding of the mechanisms through which SMM contributes to improved SCM performance. Therefore, Figure 6.4 illustrates the proposed conceptual model, positioning SMM at the intersection of strategic resources, dynamic capabilities, and transaction cost efficiencies, leading to improved supply chain sustainability and resilience.

6.5 DISCUSSIONS

The purpose of this study was to examine the role of SMM in SCM through the integration of three theoretical perspectives: the RBV, the DCV, and TCE. The findings highlight how SMM functions as a strategic resource, a capability enabler, and a cost-reduction mechanism, offering novel insights into digital transformation in supply chains.

6.5.1 Theoretical Contributions

This study advances SCM theory in three main ways. First, it extends the RBV by positioning SMM as a strategic digital resource that enhances supply chain competitiveness through improved information flow, relationship-building, and customer responsiveness. Second, it enriches the DCV by illustrating how SMM fosters adaptability, agility, and resilience in turbulent environments, particularly during crises such as the COVID-19 pandemic. Third, it applies TCE to demonstrate how SMM reduces coordination, negotiation, and monitoring costs, thereby improving efficiency and trust in supply chain partnerships. Importantly, by synthesizing these perspectives, the study addresses theoretical fragmentation in prior research and proposes an integrated conceptual framework for understanding SMM in SCM.

6.5.2 Managerial and Policy Implications

For practitioners, the findings suggest that SMM offers SMEs a cost-effective tool for achieving both efficiency and resilience. Managers can leverage social media platforms to coordinate supply chain operations, enhance visibility, and respond quickly to disruptions. By embedding SMM into supply chain strategies, firms can strengthen trust and collaboration with partners, reduce transaction costs, and improve customer service. For policymakers, the study highlights the need to support SMEs in adopting digital tools through training, infrastructure investments, and policies that encourage technology-driven supply chain innovation.

6.5.3 Limitations of the Study

This study is limited by its reliance on secondary data from existing literature rather than original empirical investigation. The findings are therefore conceptual and may not fully capture the contextual nuances of SMM adoption across different industries or regions. Additionally, the review focused on studies published between 2017 and 2025, which, while contemporary, may exclude earlier foundational research or very recent developments.

6.5.4 Future Research Directions

Future research could empirically validate the conceptual model by applying quantitative methods such as structural equation modeling (SEM) or qualitative approaches such as case studies across different SME contexts. Comparative studies between developed and developing economies would also enrich understanding of how resource availability shapes SMM adoption in SCM. Furthermore, future work could explore emerging digital platforms (e.g., TikTok, WhatsApp, Facebook, LinkedIn, Instagram, and Twitter) and technologies (e.g., AI-driven social media analytics) to assess their evolving role in supply chain resilience and sustainability.

6.6 CONCLUSION

The incorporation of SMM into SCM has altered how firms communicate, cooperate, and respond to market changes (Hartono et al., 2020). This study explored the theoretical underpinnings of this integration using the RBV, DCV, and TCE theories. These perspectives provide a deeper understanding of how firms can leverage social media to enhance supply chain efficiency, improve decision-making, and maintain competitive advantage (Kashurnikov et al., 2019; Hartono et al., 2020; Saddhono et al., 2020; Xu et al., 2021; Gao et al., 2023). From an RBV perspective, social media serves as a VRIN resource, enabling businesses to gain a competitive edge by enhancing market intelligence, real-time communication, and customer engagement (Xu et al., 2021). Companies that effectively integrate social media into their supply chains can develop unique capabilities that differentiate them from competitors, ultimately leading to greater efficiency and responsiveness (Saddhono et al., 2020). In addition, the DCV theory highlights the importance of adaptability in volatile and rapidly changing markets. Social media provides businesses with the ability to sense and respond to shifts in consumer demand, supplier availability, and industry trends. Regarding TCE, SMM reduces transaction costs, minimizes information asymmetry, and improves supply chain coordination (Lee & Zhong, 2020). By utilizing social media for supplier negotiations, logistics tracking, and real-time inventory updates, companies can lower operational expenses and streamline supply chain processes. The transparency and accessibility provided by digital platforms foster stronger supplier and consumer relationships, ensuring cost-effective and efficient supply chain operations (Gao et al., 2023).

Despite its advantages, integrating social media into SCM poses challenges, including data security risks, misinformation, and dependence on platform algorithms (Orji et al., 2020). To fully realize its benefits, firms must implement robust digital strategies, cybersecurity safeguards, and data validation practices to ensure reliable and sustainable supply chain operations (Martínez-López et al., 2022). As digital transformation advances, the integration of SMM and SCM will become increasingly critical. Future research should examine the role of emerging technologies such as blockchain, artificial intelligence, and predictive analytics in enhancing social media-enabled supply chains. Firms that adopt these technologies are likely to improve efficiency, strengthen collaboration, and sustain long-term competitiveness in global markets.

REFERENCES

Abou Kamar, M., Albadry, O. M., Sheikhelsouk, S., Ali Al-Abyadh, M. H., & Alsetoohy, O. (2023). Dynamic capabilities influence on the operational performance of hotel food supply chains: A mediation-moderation model. *Sustainability*, *15*(18). https://doi.org/10.3390/su151813562

Alkhateri, S. N., Ahmad, S., & Ali Husseini, S. (2021). Conceptual framework of the social media adoption and its impact on small and medium-sized enterprises (SMEs) service performance in UAE. *Journal of Positive School Psychology*, *2022*(3), 4737–4746. http://journalppw.com

Andrews, K. R. (1971). *The Concept of Corporate Strategy*. Homewood, IL: Dow Jones-Irwin.

Ashiru, F., Nakpodia, F., & You, J. J. (2023). Adapting emerging digital communication technologies for resilience: Evidence from Nigerian SMEs. *Annals of Operations Research*, *327*(2), 795–823. https://doi.org/10.1007/s10479-022-05049-9

Barney, J. (1991). Firm resources and sustained competitive advantage. In J. B. Barney, & D. N. Clark (Eds.), *Resource-Based Theory: Creating and Sustaining Competitive Advantage*. Oxford: Oxford University Press. https://doi.org/10.1093/oso/9780199277681.003.0003

Basri, W. S. M., & Siam, M. R. A. (2019). Social media and corporate communication antecedents of SME sustainability performance. *Journal of Economic and Administrative Sciences*, *35*(3), 172–182. https://doi.org/10.1108/jeas-01-2018-0011

Birasnav, M., & Bienstock, J. (2019). Supply chain integration, advanced manufacturing technology, and strategic leadership: An empirical study. *Computers and Industrial Engineering*, *130*, 142–157. https://doi.org/10.1016/j.cie.2019.01.021

Cao, D., Meadows, M., Wong, D., & Xia, S. (2021). Understanding consumers' social media engagement behaviour: An examination of the moderation effect of social media context. *Journal of Business Research*, *122*, 835–846. https://doi.org/10.1016/j.jbusres.2020.06.025

Chatterjee, S., Chaudhuri, R., Sakka, G., Grandhi, B., Galati, A., Siachou, E., & Vrontis, D. (2021). Adoption of social media marketing for sustainable business growth of SMEs in emerging economies: The moderating role of leadership support. *Sustainability*, *13*(21). https://doi.org/10.3390/su132112134

Cheung, M. L., Pires, G., & Rosenberger, P. J. (2020). The influence of perceived social media marketing elements on consumer–brand engagement and brand knowledge. *Asia Pacific Journal of Marketing and Logistics*, *32*(3), 695–720. https://doi.org/10.1108/APJML-04-2019-0262

Coase, R. (1937). The nature of the firm. *Economica*, *4*, 386–405.

Dubey, R., Altay, N., Gunasekaran, A., Blome, C., Papadopoulos, T., & Childe, S. J. (2018). Supply chain agility, adaptability and alignment: Empirical evidence from the Indian auto components industry. *International Journal of Operations and Production Management*, *38*(1), 129–148. https://doi.org/10.1108/IJOPM-04-2016-0173

Duong, V. X. Q., & Nguyen, H. H. (2018). Supply chain management practices, competitive advantages and firm performance: A case of small and medium enterprises (SMEs) in Vietnam. *Journal of Modern Accounting and Auditing*, *14*(3), 136–146. https://doi.org/10.17265/1548-6583/2018.03.004

Dwivedi, Y. K., Shareef, M. A., Akram, M. S., Bhatti, Z. A., & Rana, N. P. (2022). Examining the effects of enterprise social media on operational and social performance during environmental disruption. *Technological Forecasting and Social Change*, *175*, 1–35. https://doi.org/10.1016/j.techfore.2021.121364

Eisenhardt, K. M., & Martin, J. A. (2000). Dynamic capabilities: What are they? *Strategic Management Journal*, *21*(10–11), 1105–1121. https://doi.org/10.1002/1097-0266(200010/11)21:10/11<1105::AID-SMJ133>3.0.CO;2-E

Ellmarker, E. B. J. (2017). Creating a suitable strategy for innovation when the industry is frequently changing: The importance of competitive intelligence, pp. 1–12.

Fan, M., Qalati, S. A., Khan, M. A. S., Shah, S. M. M., Ramzan, M., & Khan, R. S. (2021). Effects of entrepreneurial orientation on social media adoption and SME performance: The moderating role of innovation capabilities. *PLoS ONE*, *16*, 1–24. https://doi.org/10.1371/journal.pone.0247320

Filho, V. A. V., & Moori, R. G. (2020). RBV in a context of supply chain management. *Gestao e Producao*, *27*(4), 1–20. https://doi.org/10.1590/0104-530X4731-20

Gao, J., Siddik, A. B., Khawar Abbas, S., Hamayun, M., Masukujjaman, M., & Alam, S. S. (2023). Impact of E-commerce and digital marketing adoption on the financial and sustainability performance of MSMEs during the COVID-19 pandemic: An empirical study. *Sustainability*, *15*(2). https://doi.org/10.3390/su15021594

Garfamy, R. M. (2012). Supply management: A transaction cost economics framework. *South East European Journal of Economics and Business*, *7*(2), 139–147. https://doi.org/10.2478/v10033-012-0022-6

Gavronski, I., Nascimento, L. F., & Fensterseifer, J. E. (2006). *RBV and capabilities creation in the supply chain: The case of environmental management. Seventeenth Annual POMS Conference*, 1–14.

Grant, R. M. (1991). The resource-based theory of competitive advantage: Implications for strategy formulation. *California Management Review*, *33*(3), 114–135. https://doi.org/10.2307/41166664

Grant, R. M. (1996). Prospering in dynamically – Competitive environments: Organizational capability as knowledge integration. *Knowledge and Strategy*, *7*(4), 375–387. https://doi.org/10.1016/b978-0-7506-7088-3.50011-5

Grover, V., & Malhotra, M. K. (2003). Transaction cost framework in operations and supply chain management research: Theory and measurement. *Journal of Operations Management*, *21*(4), 457–473.

Hartono, B. D., Yusuf, B. P., & Arifiati, A. (2020). Implementation of digital marketing strategies through social media marketing, supply chain management and online sales of bill chilly product. *International Journal of Supply Chain Management*, *9*(4), 348–354.

Heubeck, T. (2023). Looking back to look forward: A systematic review of and research agenda for dynamic managerial capabilities. *Management Review Quarterly*. https://doi.org/10.1007/s11301-023-00359-z

Inge, G., Jan-Benedict, S. E. M., & Nirmalya, K. (2006). Make, buy, or ally: A transaction cost theory meta-analysis. *Academy of Management Journal*, *49*(3), 519–543.

Kashurnikov, S. N., Sevalnev, V. V., Truntsevsky, Y. V., Cherepanova, E. V., & Berestneva, O. G. (2019). E-commerce in supply chain management: Its introduction and prospects in the light industry. *International Journal of Supply Chain Management*, *8*(4), 727–732.

Lee, L. S., & Zhong, W. (2020). Dependence structure, trust dimensions, and governance choices in Asian marketing channels: Evidence in China. *Asian Journal of Business Research*, *10*(1), 47–67. https://doi.org/10.14707/ajbr.200075

Madhani, P. (2010). Resource based view (RBV) of competitive advantage: An overview. In P. Madhani (Ed.), *Resource Based View: Concepts and Practices*. Hyderabad, India: Icfai University Press, 3–22. https://ssrn.com/abstract=1578704

Martínez-López, F. J., Li, Y., & Young, S. M. (2022). How companies can exploit the commercial value of social media through advertising. In F. J. Martínez-López, Y. Li, & S. M. Young (Eds.), *Social Media Monetization: Platforms, Strategic Models and Critical Success Factors*. Cham: Springer International Publishing, 13–26.

Mas Suariedewi, I. G. A. A., Adyatma, I. W. C., & Wulandari, I. G. A. A. (2022). The impact of social media marketing for MSMEs sustainability in Covid-19 period. *Journal of Business and Management Review*, *3*(9), 641–655. https://doi.org/10.47153/jbmr39.4602022

Munir, M., Jajja, M. S. S., Chatha, K. A., & Farooq, S. (2020). Supply chain risk management and operational performance: The enabling role of supply chain integration. *International Journal of Production Economics*, *227*. https://doi.org/10.1016/j.ijpe.2020.107667

Nasrollahi, M. (2018). The impact of firm's social media applications on green supply chain management. *International Journal of Supply Chain Management*, *7*(1), 16–24.

Neves, L. W. de A., Hamacher, S., & Scavard, L. F. (2014). Outsourcing from the perspectives of TCE and RBV: A multiple case study. *Producao*, *24*(3), 687–699. https://doi.org/10.1590/S0103-65132013005000082

Ninib, D., Tomeh, W., & Daniel, D. (2020). Social media marketing activities. *Business Administration*, May 2020, 1–62. https://www.diva-portal.org/smash/get/diva2:1436049/FULLTEXT01.pdf

Ojong-Ejoh, M. U., Angioha, P. U., Agba, R. U., Aniah, E. A., Salimon, M. G., & Akintola, A. (2021). Operating SMEs in the face of the Covid-19 pandemic in Calabar. *Quantitative Economics and Management Studies*, *2*(4). https://doi.org/10.35877/454ri.qems305

Ojubanire, O. A., & Popoola, D. O. (2022). Supply chain management practice among small and medium manufacturing enterprises in Nigeria: Current state and implications. *Journal of Management Info*, *8*(4), 304–318. https://doi.org/10.31580/jmi.v8i4.2342

Orji, I. J., Kusi-Sarpong, S., & Gupta, H. (2020). The critical success factors of using social media for supply chain social sustainability in the freight logistics industry. *International Journal of Production Research*, *58*(5), 1522–1539. https://doi.org/10.1080/00207543.2019.1660829

Porter, M. E. (1980). Techniques for analyzing industries and competitors. In *Competitive Strategy*. New York: Free. https://doi.org/10.1201/9781003067573

Porter, M. E. (1985). *Competitive Advantage: Creating and Sustaining Superior Performance – With a New Introduction*. New York: Free Press, 557.

Prajogo, D., & Olhager, J. (2012). Supply chain integration and performance: The effects of long-term relationships, information technology and sharing, and logistics integration. *International Journal of Production Economics*, *135*(1), 514–522. https://doi.org/10.1016/j.ijpe.2011.09.001

Ramos, E., Patrucco, A., & Chavez, M. (2023). Dynamic capabilities in the "new normal": A study of organizational flexibility, integration and agility in the Peruvian coffee supply chain. *Supply Chain Management*, *28*(1), 55–73.

Saddhono, K., Gumelar, G., Aswin, Nasrul, Alfian, F. Y., Badruddin, S., Paramitasari, N., Siregar, R., & Halim, P. (2020). Role of social media marketing to enhance the supply chain and business management. *International Journal of Supply Chain Management*, *9*(2), 1104–1107.

Schilke, O., Hu, S., & Helfat, C. E. (2018). Quo vadis, dynamic capabilities? A content-analytic review of the current state of knowledge and recommendations for future research. *Academy of Management Annals*, *12*(1), 390–439. https://doi.org/10.5465/annals.2016.0014

Sengupta, T., Narayanamurthy, G., Moser, R., Pereira, V., & Bhattacharjee, D. (2022). Disruptive technologies for achieving supply chain resilience in COVID-19 era: An implementation case study of satellite imagery and blockchain technologies in fish supply chain. *Information Systems Frontiers*, *24*(4), 1107–1123. https://doi.org/10.1007/s10796-021-10228-3

Shibin, K. T., Dubey, R., Gunasekaran, A., Hazen, B., Roubaud, D., Gupta, S., & Foropon, C. (2020). Examining sustainable supply chain management of SMEs using resource based view and institutional theory. *Annals of Operations Research*, *290*, 301–326.

Singh, S., Kumar, R., Panchal, R., & Tiwari, M. K. (2021). Impact of COVID-19 on logistics systems and disruptions in food supply chain. *International Journal of Production Research*, *59*(7), 1993–2008. https://doi.org/10.1080/00207543.2020.1792000

Snyder, H. (2019). Literature review as a research methodology: An overview and guidelines. *Journal of Business Research*, *104*, 333–339. https://doi.org/10.1016/j.jbusres.2019.07.039

Swallehe, O., & Haule, S. D. (2021). The effects of social media marketing on business performance of supermarkets in Tanzania. *Turk Turizm Arastirmalari Dergisi*, *3*(4), 234–249.

Teece, D. J. (2007). Explicating dynamic capabilities: The nature and microfoundations of (sustainable) enterprise performance. *Strategic Management Journal* *28*(13), 1319–1350. https://doi.org/10.1002/smj.640

Teece, D. J. (2018). Business models and dynamic capabilities. *Long Range Planning, 51*(1), 40–49. https://doi.org/10.1016/j.lrp.2017.06.007

Teece, D. J., Pisano, G., & Shuen, A. (1997). Dynamic capabilities and strategic management. *Knowledge and Strategy, 18*, 77–116. https://doi.org/10.1093/oso/9780198781806.003.0019

Thongrawd, C., Skulitsariyaporn, C., Sirisopana, S., & Chomchom, N. (2020). The impact of the strategic supplier partnership, and strategic outsourcing on the supply chain performance: The mediating role of customer relationship. *International Journal of Supply Chain Management, 9*(2), 562–571.

Tungande, F., Meyer, A., & Niemann, W. (2020). Opportunities and challenges of social media in supply chain management: A study in the South African FMCG retail industry. *Acta Commercii, 20*(1), 1–12. https://doi.org/10.4102/ac.v20i1.864

Uddin, M., Razzak, M., & Abd Rahman, A. (2023). Sustainable supply chain management practices, dynamic capabilities, and competitive advantage: Evidence from the Bangladesh ready-made garments industry. *Business Strategy and Development, 6*(2), 1–13. https://doi.org/10.1002/bsd2.232

Venciūtė, D. (2018). Social media marketing – From tool to capability. *Management of Organizations: Systematic Research, 79*(1), 131–145. https://doi.org/10.1515/mosr-2018-0009

Wali, A. F., Cyprian, J., & Nkpurukwe, O. I. (2020). Influences of social media marketing tools towards online purchases: Nuances from undergraduate students in Nigeria Public Universities. *Paradigm, 24*(2), 239–253.

Wernerfelt, B. (1984). The resource-based view of the firm. *Strategic Management Journal, 5*(2), 171–180.

Wijaya, O. Y. A., Sulistiyani, Pudjowati, J., Kartikawati, T. S., Kurniasih, N., & Purwanto, A. (2021). The role of social media marketing, entertainment, customization, trendiness, interaction and word-of-mouth on purchase intention: An empirical study from Indonesian smartphone consumers. *International Journal of Data and Network Science, 5*(3), 231–238. https://doi.org/10.5267/j.ijdns.2021.6.011

Williamson, O. E. (1975). Markets and hierarchies: Analysis and antitrust implications. *Administrative Science Quarterly, 22*(3), 540. https://doi.org/10.2307/2392191

Williamson, O. E. (1979). Transaction-cost economics: The governance of contractual relations. *The Journal of Law and Economics, 22*(2), 233–261. https://doi.org/10.1093/oso/9780198774358.003.0007

Williamson, O. E. (2010). Transaction cost economics: The natural progression. *Nobel Lectures: Economic Sciences: 2006–2010, 100*, 673–690. https://doi.org/10.1142/9789814635585_0004

Xu, H., & Zhao, C. (2022). Supply chain management practices influence supply chain performance with mediation role of innovation and moderation role of top management support. *Frontiers in Public Health, 10*, 1–12. https://doi.org/10.3389/fpubh.2022.813828

Xu, S., Chen, J., Wu, M., & Zhao, C. (2021). E-commerce supply chain process optimization based on whole-process sharing of internet of things identification technology. *CMES – Computer Modeling in Engineering and Sciences, 126*(2), 843–854. https://doi.org/10.32604/cmes.2021.014265

Yusriadi, Y., Awaluddin, M., Firman, H., & Asrijal, A. (2022). Implementation of e-commerce in supply chain management. *Uncertain Supply Chain Management, 10*(4), 1279–1288. https://doi.org/10.5267/j.uscm.2022.7.012

Zahoor, N., Golgeci, I., Haapanen, L., Ali, I., & Arslan, A. (2022). The role of dynamic capabilities and strategic agility of B2B high-tech small and medium-sized enterprises during COVID-19 pandemic: Exploratory case studies from Finland. *Industrial Marketing Management, 105*, 502–514. https://doi.org/10.1016/j.indmarman.2022.07.006

7 Technology-Driven Sustainability in Supply Chain Management

Himanshu Gaur, Sapna Dadwal, and Prabha Arya

7.1 INTRODUCTION

Supply-chain management has seen a significant transformation due to the growing emphasis on sustainability, which has compelled companies to include sustainability's social, economic, and environmental facets into their business plans (Elkington, 1998; Liu and Xu, 2017). Recent research contends that emerging technologies, such as artificial intelligence (AI), the Internet of Things (IoT), blockchain, big-data analytics, and robotics, are crucial enablers of sustainable supply-chain management (SSCM) in contrast to traditional supply chains, which prioritized cost effectiveness and operations effectiveness (Ben-Daya et al., 2019; Dubey et al., 2019). SSCM can be a methodical way to combine corporate operations with social responsibility, environmental stewardship, and long-term economic success in the given circumstances (Ben-Daya et al., 2019; Dubey et al., 2019).

The Triple Bottom Line (TBL) model illustrates the harmonious integration of the environment, sustainability, and economy (Elkington, 1998). Environmental, social, and economic sustainability are applicable in this construct since they react to resource conservation, fair working, ethical sourcing, business efficiency, resilience, and profitability (Ben-Daya et al., 2019; Dubey et al., 2019). The TBL sustainability is extended by the use of emerging technologies that can assist in terms of the real-time monitoring, predictive analytics, and transparent reporting throughout the supply chain.

The blockchain enhances traceability and transparency, which helps in endorsing ethical sourcing and the momentum of the circular economy (Kouhizadeh and Sarkis, 2018). Further, robotics and automation bring operational efficiency and environmentally favorable environments, which highlights technology as the key driver in grasping sustainable supply chains (Dubey et al., 2021).

Owing to the aforementioned shortcomings, it is urgently necessary to conduct a timely systematic evaluation that highlights how new technologies affect the supply chains' three sustainability pillars. Such a study is justified by the rapid speed of digitization in the field of supply chain management as well as the growing emphasis on

DOI: 10.1201/9781003770008-10

technology-enhanced sustainability in management and policy. Accordingly, this chapter begins with a systematic literature review (SLR) that uses thematic analysis to find, classify, and synthesize common themes associated with technology-based SSCM.

The specific objectives of this study are:

1. To analyze the impact of technology adoption on environmental, social, and economic sustainability in supply chains (TBL perspective).
2. To evaluate the opportunities and challenges associated with technology-driven approaches to SSCM.
3. To identify critical research gaps in the literature on technology-enabled sustainability.
4. To provide strategic directions for integrating technology and sustainability in supply chain practices.

To achieve these objectives, the study is guided by the following research questions:

1. RQ1: How do emerging technologies (AI, IoT, blockchain, big data, robotics) influence environmental, social, and economic sustainability in supply chains?
2. RQ2: What are the key opportunities and challenges of technology-driven approaches in achieving sustainable supply chain management?
3. RQ3: What are the major research gaps and future directions for strengthening the integration of technology and sustainability in SCM?

7.2 RESEARCH METHODOLOGY

7.2.1 Research Design

This research paper will use a qualitative research design that will be grounded purely on the use of thematic analysis to investigate the subject of technology-driven sustainability in supply chain management. Thematic analysis was used to determine the patterns, mechanisms, and conceptual connections that exist in previous studies in order to have an integrative picture of how digital technologies impact the outcome of environmental, social, and economic sustainability.

7.2.2 Data Source and Time Frame

The review was conducted on 150 peer-reviewed research articles that were found in the Web of Science (WoS) database only. WoS was chosen since it has stringent indexing criteria and has high coverage of high-impact journals in supply chain management, sustainability, operations management, and information systems. The literature review will discuss the 10–12 years (2012–2024) of literature due to the fast-changing nature of digital and Industry 4.0 technologies.

7.2.3 Search Keywords

Keywords were used to conduct the literature search based on such terms as sustainable supply chain management, technology-driven sustainability, Industry 4.0, artificial intelligence, Internet of Things (IoT), blockchain, big data analytics, cloud computing, and digital transformation.

7.2.4 Inclusion and Exclusion Criteria

Studies were included if they (1) focused on the application of digital technologies within supply chain contexts, (2) examined environmental, social, and/or economic sustainability outcomes, and (3) were published in English in peer-reviewed journals indexed in Web of Science. Studies were excluded if they lacked a sustainability focus, addressed technology without relevance to supply chain management, or were non-peer-reviewed or purely technical without managerial or sustainability implications.

7.2.5 Thematic Analysis Procedure

The articles that have been selected were examined by an iterative thematic analysis process that involved the process of repeated familiarization with the literature, systematic coding with the relevant concepts, and synthesis of recurring patterns. Codes associated with technology application, sustainability mechanism, outcomes, and challenges associated with adoption were then categorized as higher-order themes. Ongoing refinement provided conceptual consistency and correspondence to the aims of the study.

7.2.6 Themes Identified

The thematic analysis resulted in the identification of four dominant themes:

- Anticipatory intelligence and information-based sustainability.
- Digital infrastructures of transparency, traceability, and trust.
- Online tracking and management of resource effectiveness.
- Institutional and organizational barriers to technology-based sustainability.

These are the themes that constitute the analytical basis of the study results and justify the multidimensionality of the role of digital technologies in the development of sustainable supply chain management.

7.3 LITERATURE REVIEW

In recent years, the field of supply chain management has evolved greatly due to the increased focus on sustainability and the uptake of new technologies. The scope of sustainability in supply chains is threefold and is frequently discussed as the TBL (Elkington, 1998). Increasingly, the expectation of the firms is that the resource use will be optimized, minimum impact on the environment realized, ethical treatment of the labor met, and economic efficiency achieved in an integrated manner. This should be since this multidimensional emphasis on sustainability has become a strategic concern that affects the working practices, policies, and performance indicators of organizations in various industries.

This review takes the thematic approach where the findings of the various research studies, conceptual frameworks, and industrial practices in the area are synthesized to give a bit of holistic view of the role of technology in supporting sustainable performance of the supply chain. Through the perspective of environmental, social, and economic results that the current technology-driven sustainability literature delivers, the review is aimed at providing a well-rounded view on the matter and hence inform and navigate academic research and practical application of the topic.

7.3.1 Technology and Sustainability in Supply Chain Management

Sustainable supply chain management has also been redefined through the digital transformation and Industry 4.0 concept. Industry 4.0, characterized by a combination of cyber-physical platforms, digitalization, and smart automation, offers companies a chance to increase supply chain visibility, flexibility, and eco-efficiency in the context of a global network (Liu and Xu, 2017). Digital innovations can help businesses to more efficiently balance their operations strategies with the sustainability goals, make smarter planning decisions, and monitor the environmental and social effects in real time. In turn, this evolution is becoming an operational excellence and sustainability driver in the modern supply chains (Bag et al., 2021).

The artificial intelligence (AI), Internet of Things (IoT), blockchain, big data analytics, cloud computing, and robotics technologies are finding their way into the supply chain operations to not only improve efficiency but also make the supply chain processes more sustainable. An example would be predictive demand planning, energy planning, and waste management, which can be operated with the help of AI and big-data analytics through the advanced algorithms and machine-learning models (Wamba et al., 2018). The concept of IoT can be used to track and trace the goods and materials along the supply and production chains in real time to monitor the consumption of energy and yield of transportation and the compliance with the rules that comply with the environment-friendly principles (Ben-Daya et al., 2019).

7.3.2 Impact of Technology Adoption on Environmental, Social, and Economic Sustainability

7.3.2.1 Environmental Sustainability

Environmental sustainability in supply chain management has become increasingly dependent upon the employment of advanced technologies. Environmental sustainability is concerned with the way resources and natural resources are exploited, environmental discharge and waste minimization, and environment-friendly practices applied in all the supply chain processes which are supply, manufacturing, logistics, and distribution (Mishra et al., 2021). With responsibilities and pressures mounting on companies to monitor and improve the environmental performance of a complex chain of suppliers, companies are increasingly adopting digital technologies like big data analytics, artificial intelligence (AI), the Internet of Things (IoT), and blockchain to track, monitor, and optimize the environmental performance of a vast and complex supply network.

Big Data Analytics enables an organization to collect, process, and analyze the huge amounts of operational data and can detect and prevent operational inefficiencies and allow the organization to use the available resources more effectively. In the form of an illustration, the consumption of energy within the warehouses and manufacturing units can be analyzed in order to create initiatives to lessen the use of energy. The predictive models developed on the basis of AI can predict the change in demand, which would help the organization reduce the chances of overproduction and inventory holding, one of the most significant contributors to wastage of resources and carbon emissions (Dubey et al., 2019). Reduction of energy waste and excessive use of materials can be achieved through producing according to actual demand, which means that the manufacturing plans may vary considerably.

7.3.2.2 Social Sustainability

Social sustainability within supply chains puts an accent on human welfare, labor rights, ethical practices, and fair stakeholder involvement. Whereas environmental sustainability is concerned with efficiency of resources, social sustainability is interested in how the operations affect employees, suppliers, and communities (Treiblmaier, 2019). New technologies have also been used more and more to enhance transparency, accountability, and adherence to ethical and labor standards through the supply chains.

Digital collaboration tools and cloud systems enhance communication and interaction with supply chain stakeholders such as the suppliers, employees, and local communities. Such platforms facilitate mutual decision-making, ethical sourcing activities, and tracking of community-level effects by supply chain activities (Jabbour et al., 2020). As such, organizations can monitor the social compliance, training of employees, and the audits of suppliers through digital dashboards, which increase accountability and social responsibility.

7.3.2.3 Economic Sustainability

Big data analytics also makes the industry economically sustainable, as they provide an excellent insight into supply chain operations. Using the procurement, production, and logistics data that are analyzed at large volumes, firms will be able to identify the inefficiencies, reduce the lead times, improve the selection of their suppliers, and streamline the processes, which results in the cost reductions and improved financial performance (Wamba et al., 2018; Dubey et al., 2019). These technologies are also complemented through cloud computing and the digital software platforms that support real-time working with the supply chain partners, e.g., the exchange of information between the geographical localities. It will allow making decisions faster and more informed, reduce the number of delays, and become more responsive in the supply chain and ultimately in profits and competitiveness (Saberi et al., 2019).

This part will cover the second objective, which would be to identify and examine literature themes in technological adoption to sustainability in supply chains. Research indicates that technology will increase the efficiency of the environment, social responsibility, and economic viability (Kouhizadeh and Sarkis, 2018; Ben-Daya et al., 2019; Dubey et al., 2019). Thematic analysis shows that when several technologies are combined together, they have a synergistic effect in such dimensions. Table 7.1 represents a summary of major dimensions identified as the result of the review and forms a structural representation of how emergent technologies can influence sustainable supply chain activities.

7.3.3 Opportunities and Challenges in Technology-Driven Sustainable Supply Chains

7.3.3.1 Opportunities Identified Across the Reviewed Studies

Throughout the studies being reviewed, there are a number of substantive opportunities that are identified to seize the use of digital technologies in increasing the sustainability in the supply chains. To start with, it is a recurring fact that data-intensive solutions, especially AI, machine learning, and predictive analytics, can help to plan and control better, which translates to less wastage, fewer hiccups, and fewer stock-outs and overproduction (Dubey et al., 2019; Bag et al., 2020). By transforming operational data in great volumes and at high velocities into practical forecasts, companies can align procurement, production, and distribution with the demand cues, which reduces the inefficiencies that continually bloat both expenses and environmental impacts.

Second, transparency and verifiability are the key areas of the opportunity on which many studies have collided. Tamper-resistant audit trails of material provenance, compliance of processes, and transfer of custody are created by blockchain and distributed ledgers that are typically reinforced by IoT sensing of nodes and arcs (Saberi et al., 2019; Kouhizadeh et al., 2020). This will be the basis of sustainable sustainability statements (e.g., deforestation-free, ethically sourced), reduce opportunism in multi-tier networks, and ensure increased stakeholder trust, even by consumers, NGOs, and regulators.

7.3.3.2 Challenges Identified Across the Reviewed Studies

However, in spite of these opportunities, the literature is full of challenges that are limiting the rate and extent of technology adoption to ensure sustainability. First of all, numerous bibliographies record cost and capability challenges, which are the high prices of sensors, connectivity, data infrastructure, and advanced analytics, supplemented by a digital skills shortage that creates barriers to entry, especially in small and medium-sized suppliers embedded into global chains (Saberi et al., 2019; Bag et al., 2020). In the absence of facilitative financing, shared services, or buying-based enablement, adoption is uneven, and concentrated benefits are taken by a few.

Disparate platforms, old systems, and data standards become heterogeneous, making it more difficult to integrate end-to-end, which destroys the anticipated visibility and automation benefits (Treiblmaier, 2018). Although pilots prove to be

TABLE 7.1
Thematic Synthesis of Emerging Technologies and Sustainability Outcomes in Supply Chains

Sustainability Dimension	Technology	Key Mechanisms	Examples/ Applications	Outcomes	Trade-offs/ Challenges	References
Environmental	AI and predictive analytics	Demand forecasting, production optimization	Predictive demand planning, energy-efficient production	Reduced overproduction, lower carbon emissions, and optimized resource utilization	Requires high-quality data, implementation cost	Dubey et al. (2019), Mishra et al. (2021)
	Big data analytics	Resource monitoring, energy pattern analysis	Warehouse energy use, process optimization	Energy efficiency, waste reduction	Data privacy, high computational resources	Schoenherr (2023)
	IoT and sensor networks	Real-time monitoring of energy, water, and temperature	Smart warehouses, logistics monitoring, and predictive maintenance	Reduced energy and water consumption, lower emissions	Sensor installation cost, maintenance requirements	Ben-Daya et al. (2019)
	Blockchain	Traceability, transparency, and circular economy support	Material tracking, eco-friendly sourcing, and recycling	Responsible material use, circular supply chains	High energy consumption, infrastructure cost	Kouhizadeh et al. (2020), Saberi et al. (2019)

Social	Blockchain	Traceability, transparency, and ethical sourcing	Supplier audits, product origin verification	Reduced labor violations, enhanced accountability	High implementation cost, energy use	Kouhizadeh et al. (2020), Treiblmaier (2019)
	IoT and AI	Workplace safety monitoring, predictive maintenance	Hazard detection, workload optimization	Improved worker safety, health, and productivity	Employee privacy concerns, maintenance cost	Thottoli and Ahmed (2022)
	Digital platforms/ cloud systems	Collaboration, stakeholder engagement	Dashboards for compliance tracking, supplier engagement	Better supplier relations, ethical procurement, and community engagement	Training needs, data security risks	Jabbour et al. (2020), Treiblmaier (2019)
Economic	AI and predictive analytics	Production and logistics optimization	Optimized production schedules, route planning, and inventory management	Reduced operational costs, improved efficiency, and minimized financial risk	High-quality data requirement, implementation cost	Dev et al. (2020)
	Big data analytics	Operational insights, procurement optimization	Supplier selection, lead time reduction, and process improvement	Cost reduction, better decision-making, and enhanced profitability	Data privacy, high computational resources	Wamba et al. (2018), Dubey et al. (2019)
	Cloud computing/ digital platforms	Real-time collaboration, information sharing	Cloud dashboards, collaborative platforms	Improved responsiveness, reduced delays, and additional revenue streams	Training requirements, data security concerns	Saberi et al. (2019), Jabbour et al. (2020)
	Integrated technologies	Synergistic application of AI, IoT, Big Data, and Blockchain	Smart production, logistics, and circular supply chain systems	Holistic sustainability, long-term resilience across dimensions	High implementation cost, integration complexity	Shee et al. (2018)

valuable, inter-tier, inter-region networks are not easily scaled due to incompatible architectures and governance structures (Queiroz and Wamba, 2019).

The section considers Objective 3 to review the opportunities and challenges linked to the technology-driven methods of sustainable supply chain management. It is mentioned in the literature that emerging technologies have a variety of opportunities, such as reduction in costs, operational efficiency, increased transparency, better environmental performance, or increased social responsibility (Wamba et al., 2018; Bag et al., 2021). Nevertheless, such obstacles as high implementation costs, cybersecurity issues, the shortage of skilled labor, and complex technology integration are also noted (Kouhizadeh and Sarkis, 2018; Dubey et al., 2021). The most crucial opportunities and challenges included in Table 7.2 summarize the findings of the

TABLE 7.2
Key Enablers and Barriers to Technology-Enabled Sustainability in Supply Chains

Opportunities Identified from Prior Studies	Illustrative Sources	Challenges Identified from Prior Studies	Illustrative Sources
Data-driven planning reduces waste and improves flow performance (predictive analytics, ML, and optimization).	Dubey et al. (2019); Bag et al. (2020)	High adoption and capability costs; uneven readiness among SMEs and lower-tier suppliers.	Saberi et al. (2019), Bag et al. (2020)
End-to-end traceability and verifiable claims via blockchain + IoT strengthen trust and compliance.	Kouhizadeh et al. (2020), Saberi et al. (2019)	Interoperability and standardization gaps impede scaling and dilute visibility benefits.	Treiblmaier (2018), Queiroz and Wamba (2019)
Digitally orchestrated circularity enables reverse logistics, repair, remanufacturing, and recycling.	Queiroz et al. (2019), Saberi et al. (2019)	Cybersecurity and data-privacy risks erode willingness to share data across tiers.	Queiroz and Wamba (2019)
Cloud platforms and shared data spaces enhance collaboration, responsiveness, and ecosystem alignment.	Zhong et al. (2017), Agyabeng-Mensah et al. (2021)	Technology-related sustainability trade-offs (energy intensity of blockchain/compute).	Saberi et al. (2019)
Safety, quality, and capability spillovers (monitoring, upskilling, and supplier development).	Agyabeng-Mensah et al. (2021), Saberi et al. (2019)	Workforce displacement and social frictions without parallel reskilling and governance.	Dalenogare et al. (2018), Bag et al. (2020)

review, which allow having a clear picture and understanding of the potential and the constraints regarding the application of technology to maximize the sustainability of the supply chain.

7.3.4 Research Gaps and Future Directions

Despite of the huge recognition obtained in the field of technology-driven sustainability in managing supply chains, some research gaps are still visible through the studied literature. It has been noted repeatedly that there is not much attention paid to social sustainability outcomes as opposed to environmental and economic outcomes. Although energy efficiency, carbon footprint minimization, and cost-cutting are some of the focus areas in many studies, the social effects, including labor welfare, equitable distribution of resources, and community involvement in supply chains, are not studied as much (Bag et al., 2020; Agyabeng-Mensah et al., 2021).

Besides, the combination of more than two technologies, i.e., AI, IoT, and blockchain, remains understudied, especially in terms of synergistic and cross-border implications on sustainable performance. There are few longitudinal studies of the development of technology-enabled sustainability practices over time; how long they will last and what the systemic impacts will be are largely unknown. Further studies should employ mixed applications, incorporating qualitative information about the practice in an organization with quantitative performance pointers. In order to facilitate the development of theoretical and empirical models of technology adoption which directly influence the economic, social, and environmental performance or both, conceptual and empirical technology adoption models that are directly linked to economic, social, and environmental performance should be constructed (Agyabeng-Mensah et al., 2021). Overall, addressing these research gaps will enhance the incorporation of sustainability and technology in supply chain management and will provide meaningful information to the practitioners and legislators.

7.4 RESEARCH FINDINGS

This part presents the study results obtained out of the thematic analysis of the reviewed literature. As per the qualitative research methodology in Section 7.2, the findings utilize common patterns, links, and observations of the literature instead of duplicating the conclusions of individual researches.

Thematic analysis resulted in four prevalent and connected themes explaining how digital technologies have been leveraged to support sustainable supply chain management in general: predictive intelligence and data-driven sustainability, transparency, traceability, and trust through digital infrastructures, real-time monitoring and operational control of resource efficiency, and structural and organizational obstacles to technology-driven sustainability.

A common finding throughout the studied literature is that artificial intelligence, machine learning, and analytics in big data will act as central pillars to sustainability through enhancing anticipatory decision-making. Predictive intelligence allows organizations to make production, procurement, and logistics operations more consistent with the demand trends in actual demand. Environmentally, predictive planning will decrease overproduction, surplus stock, and unnecessary shipping, which will decrease carbon emissions and waste materials. The similar mechanisms do assist in terms of being more economical, more efficient in terms of cost, and more efficient in terms of utilizing assets and/or resilience to demand fluctuations. The synthesis implies that the gains of environmental and economic sustainability do not generate but are being generated at the same time in the form of coordination data. However, it has been suggested in the literature that the effectiveness of the predictive systems is extremely dependent upon the quality of data, interoperability, and the level of maturity of the analytics. Organizations that lack integrated data structures cannot entirely materialize predictive insights into supportable sustainability developments.

The other prevailing theme relates to the question of the institutionalization of transparency and accountability in supply chains through blockchain and distributed ledger technologies. The results prove that traceability is not only a technical feature but also a governance tool that can help to comply with environmental regulations, act ethically, and win consumer confidence. Environmental sustainability is made stronger by blockchain traces because it allows monitoring life cycle, implementing the principles of a circular economy, and checking the compliance with various regulations. The benefits of social sustainability arise in the form of better monitoring of labor standards and supplier behavior, as well as ethical sourcing commitments. Synthesis of the above shows that trust and verifiability are cross-cutting enablers, which strengthen all of the three pillars of sustainability. However, trade-offs are also identified in the findings, especially regarding the energy intensity, the cost of infrastructure, and the issue of scalability. In this way, the positive role of transparency-enhancing technologies in enhancing sustainability is only attainable in the case of responsible system design and governance structures (Table 7.3).

TABLE 7.3
Synthesized Research Findings on Technology-Driven Sustainability in Supply Chain Management

Sustainability Dimension	Key Technologies	Synthesized Findings	Dominant Impacts	Key Constraints
Environmental	AI, big data, IoT, blockchain	Predictive planning and real-time monitoring reduce emissions, waste, and resource inefficiency	Lower carbon footprint, circular economy enablement	Energy intensity, data integration challenges
Social	Blockchain, IoT, digital platforms	Transparency and monitoring improve labor compliance, safety, and ethical sourcing	Enhanced accountability, safer workplaces	Privacy concerns, uneven adoption

(Continued)

TABLE 7.3 (Continued)

Sustainability Dimension	Key Technologies	Synthesized Findings	Dominant Impacts	Key Constraints
Economic	AI, big data, cloud platforms	Optimization and collaboration improve efficiency, resilience, and profitability	Cost reduction, long-term competitiveness	High initial investment, skills gap
Cross-dimensional	Integrated digital ecosystems	Synergistic use of multiple technologies delivers holistic sustainability outcomes	Balanced TBL performance	Integration complexity

The findings that have been synthesized support the fact that sustainability based on technology, which is placed on the aspect of management of supply chains, is multidimensional, interdependent, and sensitive to context. Digital technologies are both powerful facilitators of sustainable performance when employed strategically and governed ethically, but they have structural constraints, trade-offs with all spirators, and contextual differences that limit their universal application.

7.5 DISCUSSION

The goal of this part is to explain the aggregate research in relation to existing theories, the background empirical studies, and factual realities related to supply chain management. Although the literature review is merely a record of the established knowledge, this discussion clarifies the causal basis of the observed patterns, their theoretical resemblance, and their practical consequences to the managerial decision-making processes. The presentation of the discussion is based on themes that are in direct agreement with the four major findings in Section 7.4.

The results show that the TBL framework (Elkington, 1998) synthesized with Dynamic Capabilities Theory (Teece et al., 1997) is the best way of explaining technology-driven sustainability in supply chain management. The available literature unanimously proves that digital technologies do not provide proportionate environmental, social, and economic returns unless the organizations are ready to develop sensing, seizing, and reconfiguring opportunities.

Systems based on artificial intelligence and analytics mainly contribute to improved economic and environmental performance by improving demand prediction, optimization of inventory, and energy efficiency. These results support the idea of Dynamic Capabilities Theory, which states that the competitive advantage is formed as the ability of an organization to rearrange operational resources in order to adapt to the complexity of the environment. Conversely, longer traditions and beliefs provide blockchain and transparency-oriented technologies with a close relationship to the institutional theory because they occur to implement them due to regulatory

TABLE 7.4
Comparison of Synthesized Findings with Prior Literature

Key Theme	Findings from This Study	Alignment with Past Literature	Contribution
Predictive analytics	Strong link to environmental and economic sustainability	Dubey et al. (2019), Wamba et al. (2018)	Confirms predictive intelligence as dual-benefit mechanism
Blockchain transparency	Strong social and governance impact	Kouhizadeh and Sarkis (2018), Saberi et al. (2019)	Extends role to legitimacy and trust-building
IoT monitoring	Operational sustainability enabler	Ben-Daya et al. (2019)	Highlights real-time control as proactive sustainability tool
Integrated technologies	Synergistic sustainability outcomes	Queiroz et al. (2019)	Strengthens argument for ecosystem-level adoption

pressures, stakeholder scrutiny, and warrant-seeking behavior rather than direct cost-cutting incentives.

Comparative synthesis shows a high degree of consistency between the current results and the previous empirical research, which narrows down their inferences at the same time. Overall, this alignment is contained in Table 7.4.

The evidence shows that the most significant impact is the increase in environmental performance in the case of the combination of predictive technologies and real-time technologies. Forecasting powered by AI also decreases production, whereas IoT sensors decrease the amount of energy loss and waste. This is in parallel to the Natural Resource-Based View (RBV) (Hart, 1995) that states that prevention of pollution and economic efficiency of resources are the origins of competitive advantage.

The results of social sustainability are still relatively underreported in the literature, which validates one of the research gaps. In case they are considered, traceability and safety monitoring systems with blockchain and IoT enhance compliance in labor and on-site health. Such results are attainable based on the stakeholder theory, which puts emphasis on the responsibility in relation to workers, communities, and civil society.

The dimension that has most continually been supported in the literature is economic sustainability. The use of AI, big data analytics, and cloud platforms facilitates the reduction of costs, minimization of risk, and responsiveness. These findings are correlated to the premises of the RBV logic, i.e., the data and analytical capabilities served as strategic resources.

First, **managers should adopt a capability-based approach to technology implementation and sustainability initiatives**. The financial statements of AI or blockchain should seek to include controls over data, worker training, and cross-partner integration models.

TABLE 7.5
Analytical Synthesis of Findings, Theory, and Implications

Theme	Theoretical Interpretation	Managerial Insight
Predictive intelligence	Dynamic capabilities theory	Invest in analytics maturity, not tools alone
Transparency and traceability	Institutional and stakeholder theory	Use blockchain to build legitimacy and trust
Real-time monitoring	Natural RBV	Integrate IoT with sustainability KPIs
Adoption barriers	Sociotechnical systems	Enable SMEs through ecosystem governance

Secondly, sustainability managers must think in terms of portfolio-based technology, with particular technologies that are correlated to desired sustainability metrics (e.g., AI to efficiency, blockchain to accountability, IoT to monitoring) derived. This prevents the creation of unrealistic hopes of single-technology solutions.

Third, when it comes to lower-tier suppliers, the lead firms should have a coordinating role in supplier enablement. Common digital platforms, funding structures, and training systems are essential in the dispersion of sustainability advantages throughout the supply chain.

In order to bring the discussion together, Table 7.5 provides an analytical summary to connect the findings, theory, and managerial relevance.

7.6 CONCLUSION

This chapter has discussed how the new digital technologies can foster sustainability in supply chain management at the environmental, social, and economic levels. However, even though the sustainable supply chains and Industry 4.0 technologies were studied rather thoroughly before, the current literature is still disjointed and tends to talk about individual tools or the results. This study synthesized using the qualitative thematic literature review to come up with an integrated literature on the topic of technology-driven sustainability in supply chains.

The results also suggest that digital technologies help achieve sustainability in varied aspects. The three main processes that can improve environmental and economic sustainability are artificial intelligence and big data analytics that increase the prediction of demand and waste reduction and optimization of the utilization of resources. The presence of blockchain-based systems will enhance social sustainability through increased transparency, traceability, and ethical accountability, whereas the IoT technologies will help monitor the usage of energy, emissions, and workplace conditions in real time. Synthesis also reveals that a combined approach of applying several technologies yields greater sustainability results compared to solitary application, though cost factors, skills, and interoperability issues are also major challenges.

Practically, the study recommends that managers be advised to have a capability-based approach toward digital sustainability initiatives, where a particular technology is being linked to a particular sustainability goal and made viable when they support it with data governance, workforce learning, and integration between organizations. Greater sustainability impacts can be achieved by supporting the transfer of technology to the upper or lower ends of the supply chain (especially unsourced suppliers) and in ethical governance, to both data privacy and surveillance of the workforce.

The limitation of the study is that it is based on secondary literature and qualitative integration, limiting the ability to cause and siphon potential bias in publication. Moreover, contextual generalizability is restricted by the prevalence of the research of the developed economies. The research in the future should follow quantitative and mixed-method research, analyze the results of social sustainability more comprehensively, and consider the long-term and synergistic consequences of integrated digital technologies in various institutional settings.

REFERENCES

Agyabeng-Mensah, Y., Afum, E., Acquah, I. S. K., Dacosta, E., Baah, C., & Ahenkorah, E. (2021). The role of green logistics management practices, supply chain traceability and logistics ecocentricity in sustainability performance. *The International Journal of Logistics Management*, 32(2), 538–566. https://doi.org/10.1108/IJLM-05-2020-0187

Bag, S., Gupta, S., Kumar, S., & Sivarajah, U. (2021). Role of technological drivers in circular economy adoption: A study on the Indian automobile sector. *Technological Forecasting and Social Change*, 165, 120557. https://doi.org/10.1016/j.techfore.2020.120557

Bag, S., Wood, L. C., Xu, L., Dhamija, P., & Kayikci, Y. (2020). Big data analytics as an operational excellence approach to enhance sustainable supply chain performance. *Resources, Conservation and Recycling*, 153, 104559. https://doi.org/10.1016/j.resconrec.2019.104559

Ben-Daya, M., Hassini, E., & Bahroun, Z. (2019). Internet of things and supply chain management: A literature review. *International Journal of Production Research*, 57(15–16), 4719–4742. https://doi.org/10.1080/00207543.2017.1402140

Dalenogare, L. S., Benitez, G. B., Ayala, N. F., & Frank, A. G. (2018). The expected contribution of Industry 4.0 technologies for industrial performance. *International Journal of production economics*, 204, 383–394. https://doi.org/10.1016/j.ijpe.2018.08.019

Dev, N. K., Shankar, R., & Qaiser, F. H. (2020). Industry 4.0 and circular economy: Operational excellence for sustainable reverse supply chain performance. *Resources, Conservation and Recycling*, 153, 104583. https://doi.org/10.1016/j.resconrec.2019.104583

Dubey, R., Gunasekaran, A., Childe, S. J., Blome, C., & Papadopoulos, T. (2019). Big data and predictive analytics and manufacturing performance: Integrating institutional theory, resource-based view and big data culture. *British Journal of Management*, 30(2), 341–361. https://doi.org/10.1111/1467-8551.12355

Dubey, R., Gunasekaran, A., Childe, S. J., Papadopoulos, T., Luo, Z., Wamba, S. F., & Roubaud, D. (2021). Blockchain technology for enhancing swift-trust, collaboration and resilience within a humanitarian supply chain setting. *International Journal of Production Research*, 59(11), 3389–3409. https://doi.org/10.1080/00207543.2020.1722860

Elkington, J. (1998). *Cannibals with Forks: The Triple Bottom Line of 21st Century Business.* New Society Publishers.

Hart, S. L. (1995). A natural-resource-based view of the firm. *Academy of Management Review*, 20(4), 986–1014. https://doi.org/10.5465/amr.1995.9512280033

Jabbour, C. J. C., de Sousa Jabbour, A. B. L., Foropon, C., & Godinho Filho, M. (2020). When titans meet–Can industry 4.0 revolutionize the environmentally-sustainable manufacturing wave? *Technological Forecasting and Social Change*, 132, 18–25. https://doi.org/10.1016/j.techfore.2018.01.017

Kouhizadeh, M., & Sarkis, J. (2018). Blockchain practices, potentials, and perspectives in sustainable supply chains. *Sustainability*, 10(10), 3652. https://doi.org/10.3390/su10103652

Kouhizadeh, M., Saberi, S., & Sarkis, J. (2020). Blockchain technology and the sustainable supply chain: Theoretically exploring adoption barriers. *International Journal of Production Economics*, 231, 107831. https://doi.org/10.1016/j.ijpe.2020.107831

Liu, Y., & Xu, X. (2017). Industry 4.0 and cloud manufacturing: A comparative analysis. *Journal of Manufacturing Science and Engineering*, 139(3), 034701. https://doi.org/10.1115/1.4034667

Mishra, D., Gunasekaran, A., & Papadopoulos, T. (2021). Big data and supply chain management: A review and bibliometric analysis. *Annals of Operations Research*, 270(1–2), 313–336. https://doi.org/10.1007/s10479-016-2236-y

Queiroz, M. M., & Wamba, S. F. (2019). Blockchain adoption challenges in supply chain: An empirical investigation of the main drivers in India and the USA. *International Journal of Information Management*, 46, 70–82. https://doi.org/10.1016/j.ijinfomgt.2018.11.021.

Saberi, S., Kouhizadeh, M., Sarkis, J., & Shen, L. (2019). Blockchain technology and its relationships to sustainable supply chain management. *International Journal of Production Research*, 57(7), 2117–2135. https://doi.org/10.1080/00207543.2018.1533261

Schoenherr, T. (2023). Supply chain management professionals' proficiency in big data analytics: Antecedents and impact on performance. *Transportation Research Part E: Logistics and Transportation Review*, 169, 102972. https://doi.org/10.1016/j.tre.2022.102972

Shee, H., Miah, S. J., Fairfield, L., & Pujawan, I. N. (2018). The impact of cloud-enabled process integration on supply chain performance and firm sustainability: The moderating role of top management. *Supply Chain Management: An International Journal*, 23(6), 500–517.

Teece, D. J., Pisano, G., & Shuen, A. (1997). Dynamic capabilities and strategic management. *Strategic Management Journal*, 18(7), 509–533. https://doi.org/10.1002/(SICI)1097-0266(199708)18:7%3C509::AID-SMJ882%3E3.0.CO;2-Z

Thottoli, M. M., & Ahmed, E. R. (2022). Information technology and e-accounting: Some determinants among SMEs. *Journal of Money and Business*, 2(1), 1–15.

Treiblmaier, H. (2018). The impact of the blockchain on the supply chain: a theory-based research framework and a call for action. *Supply Chain Management: An International Journal*, 23(6), 545–559. https://doi.org/10.1108/SCM-01-2018-0029

Treiblmaier, H. (2019). Combining blockchain technology and the physical internet to achieve triple bottom line sustainability: A comprehensive research agenda for modern logistics and supply chain management. *Logistics*, 3(1), 10. https://doi.org/10.3390/logistics3010010

Wamba, S. F., Akter, S., Edwards, A., Chopin, G., & Gnanzou, D. (2018). How 'big data' can make big impact: Findings from a systematic review and a longitudinal case study. *International Journal of Production Economics*, 165, 234–246. https://doi.org/10.1016/j.ijpe.2014.12.031

Zhong, R. Y., Xu, X., Klotz, E., & Newman, S. T. (2017). Intelligent manufacturing in the context of industry 4.0: A review. *Engineering*, 3(5), 616–630. https://doi.org/10.1016/J.ENG.2017.05.015

Section IV

Circular Economy and Reverse Logistics

8 Exploring the Synergy of Green Manufacturing, Sustainable Supply Chains, and Circular Economy

A Systematic Review of Strategies, Challenges, and Opportunities

Mallika Roy, Nabila Maruf, and Abul Hasnat Muhammed Salimullah

8.1 INTRODUCTION

The urgency of today's environmental pressures has increasingly pushed industries to rethink how production and supply systems operate (Blok et al., 2015). A substantial body of research shows that adopting greener manufacturing (GM) practices can support economic performance while reducing environmental impacts (Pimenov et al., 2022). Recent studies further stress that sustainability outcomes depend not only on cleaner production methods but also on responsible resource use across supply chains, particularly during production and distribution stages (Khandare, 2024; Vangeri et al., 2024). Related work by Roy and Akbar (2025) extends this discussion by examining pro-environmental behaviour in agricultural supply chains, with particular emphasis on food waste reduction through improved handling practices, consumer awareness, and food recovery initiatives. This study integrates research on circular business models, sustainable manufacturing, and supply chains to show how their interaction supports a gradual shift towards sustainability. While GM improves resource efficiency and green supply chains extend sustainability across

DOI: 10.1201/9781003770008-12

operations, alignment with CE principles enhances waste reduction, material reuse, and long-term economic and technological performance (Patil et al., 2024; Roy, 2025; Roy et al., 2025a; Vangeri et al., 2024). Building on prior research, this study examines how GM, SSC, and CE practices jointly shape environmental, economic, and technological outcomes.

8.2 LITERATURE SYNTHESIS

Mounting environmental pressures are driving industries towards GM, which improves resource efficiency while supporting environmental performance and sectoral growth (Goffi et al., 2018; Wannags & Gold, 2020; Sustainable Manufacturing Report, 2025). Sustainability extends beyond factory boundaries through environmentally responsible procurement, production, logistics, and distribution, while the CE complements this approach by promoting longer product use, waste reduction, and improved material recovery (Grant et al., 2017; Elisha, 2022). Consumer preferences increasingly favour sustainability, with 72% globally and 78% in the United States willing to pay for environmentally responsible products (Arbor, 2025). Although research on GM, SSC, and CE is growing, these areas are often studied separately, limiting insights into their combined impact. Figure 8.1 highlights strong conceptual linkages but also reveals the need to better connect theory with industry-level practice.

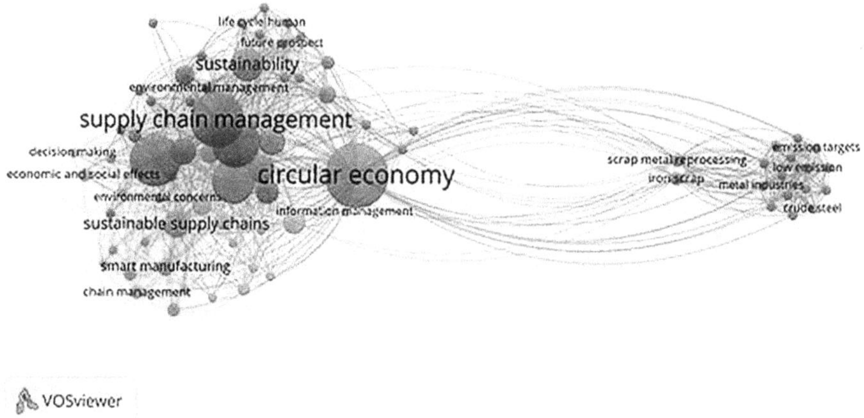

FIGURE 8.1 Keyword co-occurrences by all keywords in Scopus database.

(Generated by authors.)

8.3 METHODOLOGY

This systematic review follows a structured and transparent approach designed to ensure a comprehensive and balanced selection of relevant studies.

8.3.1 Research Scope and Objectives

This review explores how sustainable manufacturing, green supply chains, and CE concepts are linked in existing research, while highlighting areas where stronger integration is still needed. To support this analysis, the review was guided by the following secondary research questions:

1. Which strategies and practices are most frequently discussed in relation to sustainable or GM?
2. How do green supply chain initiatives support and strengthen sustainable manufacturing practices?
3. How do CE principles and business models strengthen sustainability across these areas?

8.3.2 Search Strategy, Study Selection, Data Extraction, and Synthesis Process

A comprehensive search across Scopus, Web of Science, and Google Scholar was undertaken to ensure the robustness and credibility of the review. The search string combined keywords relating to sustainable manufacturing, eco-friendly supply chains, and the CE. The Boolean keywords were ("sustainable manufacturing" OR "green manufacturing" OR "eco-friendly manufacturing") AND ("sustainable supply chain" OR "green supply chain" OR "environmentally conscious supply chain" OR "eco supply chain" OR "supply chain sustainability") AND ("circular economy" OR "circular business model" OR "closed-loop system" OR "resource efficiency" OR "waste minimization"). The review focused on English-language studies published between 2000 and 2024, using peer-reviewed sources to ensure broad coverage and academic quality. Relevant studies were selected based on clear inclusion criteria focusing on GM, SSC, and CE practices, while non-peer-reviewed or methodologically weak studies were excluded.

Figure 8.2 illustrates the PRISMA flow diagram outlining the study selection process. An initial search identified 2,798 records, with 52 additional sources obtained elsewhere. After removing duplicates and irrelevant entries, 257 full-text articles were reviewed. Of these, 206 were excluded due to limited relevance or lack of peer review, leaving 51 studies that met all inclusion criteria and formed the final dataset for analysis.

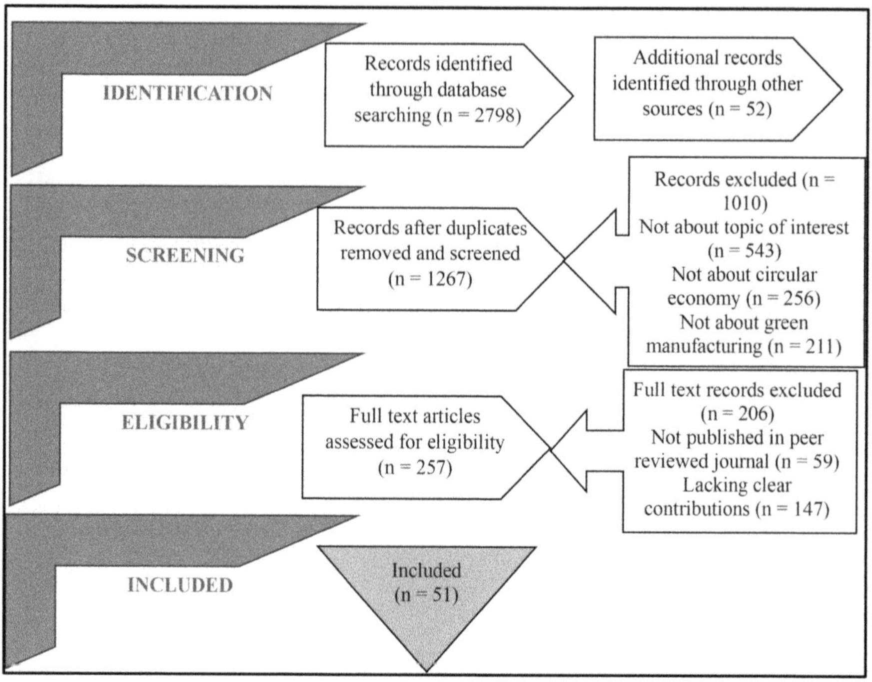

FIGURE 8.2 PRISMA diagram.

(Generated by authors.)

8.4 THEORETICAL FOUNDATION

This section outlines the key theoretical foundations underpinning GM, SSC, and CE practices, with stakeholder and institutional theories offering particular relevance. Stakeholder theory explains how firms respond to pressures from suppliers, customers, regulators, and society to adopt environmentally responsible practices and improve resource efficiency (Vachon & Klassen, 2006; Zhu et al., 2010; Crane & Ruebottom, 2011; Hsu et al., 2013; Miles, 2017; Dmytriyev et al., 2021; Baah et al., 2022). Institutional theory further highlights how regulatory, normative, and cultural forces shape organizational behaviour and encourage sustainability adoption through policies, standards, and social expectations (Zucker, 1987; Jennings & Zandbergen, 1995; Escobar & Vredenburg, 2011; Brammer et al., 2012; Chiu, 2018; Risi et al., 2023).

8.5 OVERVIEW OF THE PUBLICATIONS

8.5.1 Evolution of the Studies

Publication on a specific topic changes over time. Figure 8.3 demonstrates the evolution of the studies.

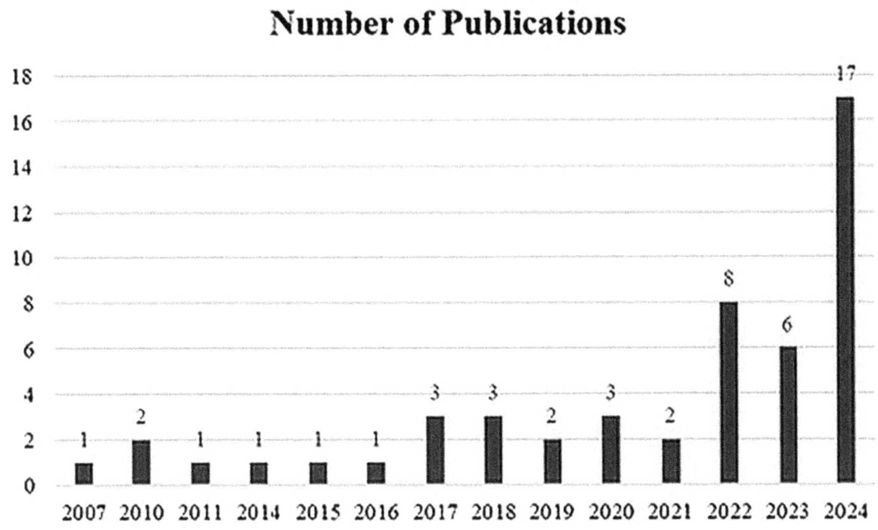

FIGURE 8.3 Evolution of the research outputs.

(Generated by authors.)

Publication trends show steady growth from 2007 to 2024. Early output remained limited, but activity increased notably after 2017. A sharp rise occurred from 2022 onwards, peaking in 2024, indicating growing academic interest, stronger collaboration, and increased research engagement in this field.

8.5.2 Type of Studies and Publication Patterns

Identifying study types helps assess the strength and balance of the evidence base.

As shown in Figure 8.4, most studies are empirical, reflecting a strong emphasis on data-driven analysis. Review papers also feature prominently, while fewer studies adopt case-based or conceptual approaches. Overall, the distribution indicates an active research field with growing analytical depth.

8.5.3 Research Approach and Country-wise Contributions

Figure 8.5 shows a balanced use of research designs, with qualitative studies dominating, followed by quantitative and mixed-method approaches. This spread reflects scholars' use of diverse methods to capture complex issues and strengthen the depth and reliability of findings.

The geographical spread of publications points to wide international participation, with a noticeable concentration in several major research centres (Figure 8.6). Research output is led by the United States, followed by China, the United Kingdom, and India, with additional contributions from several other countries. A large share of studies involves cross-national collaboration, highlighting the global and increasingly interconnected nature of research in this field.

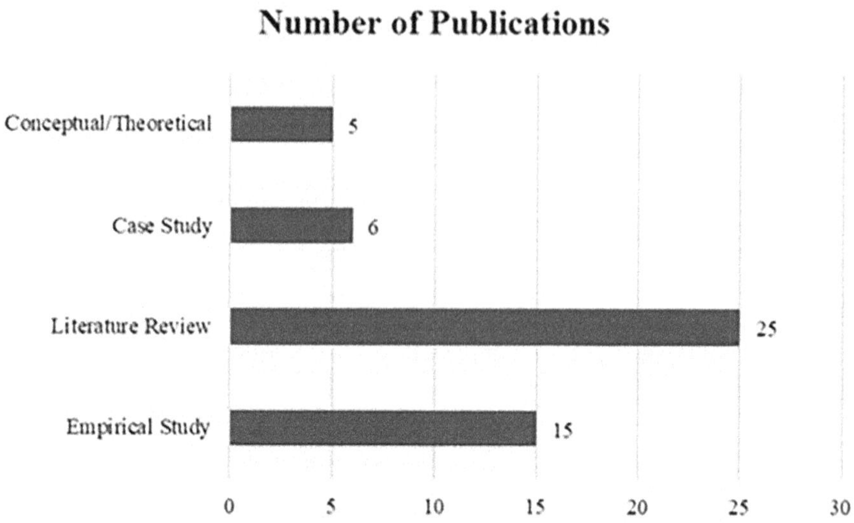

FIGURE 8.4 Type of study and publication record.

(Generated by author.)

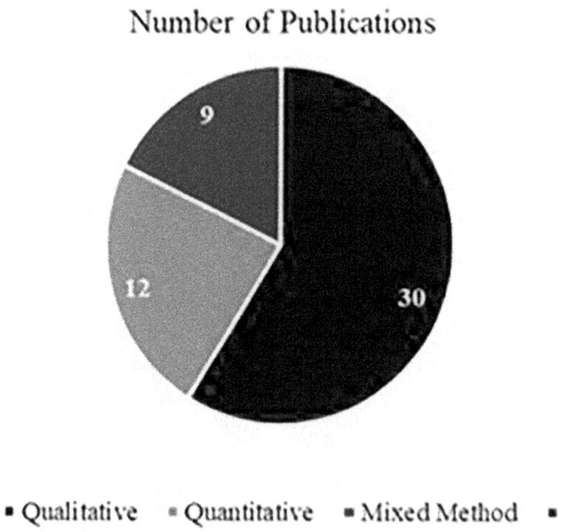

FIGURE 8.5 Research techniques and academic publications.

(Generated by authors.)

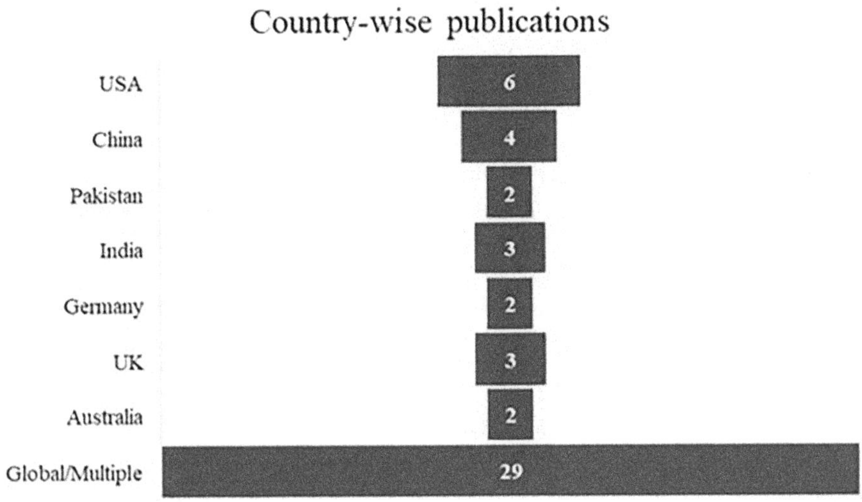

FIGURE 8.6 Countries and scholarly work.

(Generated by authors.)

8.5.4 Themes and Research Output

The thematic spread of publications indicates a strong emphasis on sustainability and innovation across supply chain and environmental research domains (Table 8.1).

The literature is largely centred on CE themes, followed by strong attention to green supply chain management (GSCM) and waste reduction. Emerging areas such as digitalization, closed-loop systems, and technological innovation further reflect growing interest in sustainability-driven transformation. Overall, the studies highlight an active and evolving research landscape focused on sustainable production and supply networks.

TABLE 8.1
Themes and Research Output

Theme	Number of Publications
Green supply chain management (GSCM)	10
Circular economy (CE)	12
Sustainable waste management	6
Digital transformation in supply chain	5
Closed-loop systems	4
Technological innovation	3
Total	49

Source: Generated by authors.

8.6 FINDINGS

8.6.1 Integration of Green Manufacturing and Supply Chains

This section examines how sustainable manufacturing aligns with supply chain activities, particularly through improved resource efficiency, waste reduction, and technological integration. Supply chain management functions as a coordinated network linking sourcing, production, logistics, and delivery, influencing multiple stakeholders (Zhang et al., 2023). Originally defined by Oliver and Webber (1982), SCM has evolved into a dynamic system shaped by planning, procurement, production, and performance evaluation (Habib, 2010; Karim et al., 2024). Growing environmental pressures have driven the shift towards GSCM, which integrates environmental considerations across all stages of the supply chain (Srivastava, 2007; Geng et al., 2017). Empirical studies confirm that GSCM improves environmental and financial performance through green purchasing, eco-design, and collaboration (Chin et al., 2015; Jo & Kwon, 2021; Ahmad et al., 2022). Recent research further highlights the role of digitalization in enhancing green performance by improving energy efficiency and enabling smarter supply chain operations (Lerman et al., 2022; Wang et al., 2020).

8.6.2 Role of Circular Economy

This section examines the application of CE principles within manufacturing and supply chains, emphasizing closed-loop systems, lifecycle management, and the shift from linear "take–make–dispose" models (EMF, 2023; UNEP, 2024). CE promotes resource efficiency, waste reduction, and extended product use through reuse, recycling, and recovery practices (Kirchherr et al., 2023; Zahran, 2024). Scholars highlight its role in reducing environmental impacts, improving resource productivity, and supporting sustainable growth (Kara et al., 2022; Karim et al., 2024). Integrating CE with supply chain management enhances performance through material reuse, digitalization, and circular logistics (Zhang et al., 2023; Farooque et al., 2019). Empirical studies further show that CE adoption strengthens sustainability, resilience, and economic outcomes across sectors, including agri-food systems (Roy & Sarker, 2025; Roy et al., 2025b).

8.6.3 Case Studies and Practical Applications

A set of case studies were examined to illustrate how GM, SSC, and CE approaches are applied in practice. From these cases, key best practices and practical lessons were derived.

8.6.3.1 Green Manufacturing: Tesla's Sustainable Production Process

Tesla is often highlighted for advancing sustainable manufacturing through its focus on electric mobility and the use of renewable energy in production (Maradin et al., 2022). The Nevada Gigafactory demonstrates how large-scale facilities can integrate sustainability into operations (Hetzel, 2021). By moving beyond traditional recycling towards repair, reuse, and refurbishment, Tesla illustrates how environmental responsibility can also deliver long-term strategic and competitive advantages (Dornfeld, 2010; Haleem et al., 2023).

8.6.3.2 Green Supply Chain: Unilever's Sustainable Sourcing

Unilever's Lipton Tea initiative illustrates how sustainability can be embedded through certified sourcing and long-term strategic commitment (Henderson & Nellemann, 2011; Murphy & Murphy, 2018). Supported by strong leadership, the company aligned operational practices with broader sustainability goals (Ionescu-Somers & Seifert, 2014). However, implementation challenges remain, particularly in developing markets, where supply chain constraints and varying consumer willingness to pay continue to limit the wider adoption of sustainable practices (Henderson & Nellemann, 2011; Almutairi, 2023).

8.6.3.3 Circular Economy: IKEA's Furniture Take-Back and Recycling Programme

CE practices are increasingly visible in the furniture sector, with IKEA's take-back and recycling initiatives often cited as leading examples. While consumers show willingness to support such models, challenges remain, particularly in knowledge gaps and implementation (Tari & Trudel, 2024). Tools such as configurators have been proposed to help firms better assess and overcome barriers to effective take-back systems (Koszewska & Bielecki, 2020; Bocken & Konietzko, 2022; Mallick et al., 2022).

8.6.3.4 Green Manufacturing and Circular Economy: Patagonia's Worn Wear Initiative

Patagonia's Worn Wear programme is widely recognized as a strong example of CE practice in the fashion industry, as it encourages repair, resale, and extended product use to reduce waste (Bürklin, 2019; Park, 2020). By shifting away from the traditional "take–make–dispose" model, the initiative supports economic and social value creation while also strengthening brand reputation and consumer trust (Rattalino, 2018; Radhakrishnan, 2021).

8.6.3.5 Circular Economy and Green Supply Chain: Renault's Car Part Recycling Initiative

The core aim of CE and GSCM is to minimize waste and improve resource efficiency (Liu et al., 2018), with the automotive sector applying circular supply chains through recycling of carbon fibre composites (Omair et al., 2022). Circular systems emphasize leasing models, localized loops, cross-sector collaboration, and lifecycle-based remanufacturing, reuse, and recycling (Dora et al., 2016; De Angelis et al., 2018).

8.6.3.6 Green Supply Chain and Circular Economy: Nike's Move to Zero Strategy

Integrating CE principles into GSCM is increasingly seen as a viable route to sustainability and waste minimization through resource recovery and reuse (Solomon et al., 2024). While GSCM and CE differ conceptually, their integration offers strong potential (Roy & Medhekar, 2024). Three-tier circular supply structures have demonstrated significant waste reduction benefits (Iqbal et al., 2020; Groenewald, 2024).

8.7 CONCLUDING REMARKS

This systematic review examined how GM, SSC, and CE principles interact, revealing both their complementarities and the challenges that shape their application. The results indicate that GM contributes significantly to reducing environmental impacts, while the CE offers a wider, systems-oriented approach focused on material reuse, waste minimization, and efficient resource use. However, the review also identifies ongoing challenges, such as technological constraints, uneven uptake across sectors, and limited coordination among key stakeholders.

REFERENCES

Ahmad, A., Ikram, A., Rehan, M. F., & Ahmad, A. (2022). Going green: Impact of green supply chain management practices on sustainability performance. *Frontiers in Psychology, 13*, 973676.

Almutairi, Y. M. (2023). Sustainability of supply chain Unilever case study. *AJME Journal, 7*(4), 210–229. https://ajme.journals.ekb.eg/article_258370.html

Arbor (2025). *80+ Sustainability Statistics for 2025: Ultimate List*. https://www.arbor.eco/blog/sustainability-statistics (accessed on 15 March 2026).

Baah, C., Afum, E., Agyabeng-Mensah, Y., & Agyeman, D. O. (2022). Stakeholder influence on adoption of circular economy principles: measuring implications for satisfaction and green legitimacy. *Circular Economy and Sustainability, 2*, 1–21.

Blok, V., Long, T. B., Gaziulusoy, A. I., Ciliz, N., Lozano, R., Huisingh, D., … & Boks, C. (2015). From best practices to bridges for a more sustainable future: Advances and challenges in the transition to global sustainable production and consumption: Introduction to the ERSCP stream of the Special volume. *Journal of Cleaner Production, 108*, 19–30.

Bocken, N., & Konietzko, J. (2022). Circular business model innovation in consumer-facing corporations. *Technological Forecasting and Social Change, 185*, 122076.

Brammer, S., Jackson, G., & Matten, D. (2012). Corporate social responsibility and institutional theory: New perspectives on private governance. *Socio-Economic Review, 10*(1), 3–28.

Bürklin, N. (2019). Worn wear: Better than new—How Patagonia's social marketing campaign enhances consumers' responsible behavior. In D. Z. Basil, G. Diaz-Meneses, & M. D. Basil (Eds.), *Social Marketing in Action: Cases from Around the World* (187–201). Springer.

Chin, T. A., Tat, H. H., & Sulaiman, Z. (2015). Green supply chain management, environmental collaboration and sustainability performance. *Procedia Cirp, 26*, 695–699.

Chiu, I. H. (2018). An institutional theory of corporate regulation. *Current Legal Problems, 71*(1), 279–334.

Crane, A., & Ruebottom, T. (2011). Stakeholder theory and social identity: Rethinking stakeholder identification. *Journal of Business Ethics, 102*, 77–87.

De Angelis, R., Howard, M., & Miemczyk, J. (2018). Supply chain management and the circular economy: Towards the circular supply chain. *Production Planning & Control, 29*(6), 425–437.

Dmytriyev, S. D., Freeman, R. E., & Hörisch, J. (2021). The relationship between stakeholder theory and corporate social responsibility: Differences, similarities, and implications for social issues in management. *Journal of Management Studies, 58*(6), 1441–1470.

Dora, M., Bhatia, M. S., & Gallear, D. (2016). Supply chain in a circular economy: A multidimensional research agenda. Working Paper, Brunel Business School.

Dornfeld, D. (2010). *Sustainable Manufacturing–Greening Processes, Systems and Products*. https://escholarship.org/content/qt80x443hk/qt80x443hk.pdf (accessed on 3 March 2026).

Elisha, O. D. (2022). Natural resources and environmental security for sustaining peacebuilding in Nigeria. *International Journal of Institutional Leadership, Policy and Management*, *4*(1), 15–38.

Ellen MacArthur Foundation (2023). *What is the Linear Economy?*. Accessed from https://www.ellenmacarthurfoundation.org/what-is-the-linear-economy (accessed 9 February 2024).

Escobar, L. F., & Vredenburg, H. (2011). Multinational oil companies and the adoption of sustainable development: A resource-based and institutional theory interpretation of adoption heterogeneity. *Journal of Business Ethics*, *98*, 39–65.

Farooque, M., Zhang, A., Thürer, M., Qu, T., & Huisingh, D. (2019). Circular supply chain management: A definition and structured literature review. *Journal of Cleaner Production*, *228*, 882–900.

Geng, R., Mansouri, S. A., & Aktas, E. (2017). The relationship between green supply chain management and performance: A meta-analysis of empirical evidences in Asian emerging economies. *International Journal of Production Economics*, *183*, 245–258.

Goffi, G., Masiero, L., & Pencarelli, T. (2018). Rethinking sustainability in the tour-operating industry: Worldwide survey of current attitudes and behaviors. *Journal of Cleaner Production*, *183*, 172–182.

Grant, D. B., Wong, C. Y., & Trautrims, A. (2017). *Sustainable Logistics and Supply Chain Management: Principles and Practices for Sustainable Operations and Management.* Kogan Page Publishers.

Groenewald, E. S. (2024). Circular economy strategies in supply chain management: Towards zero waste. *Power System Technology*, *48*(1), 464–480.

Habib, M. M. (2010). Supply chain management: Theory and its future perspectives. *International Journal of Business, Management and Social Sciences*, *1*(1), 79–87.

Haleem, A., Javaid, M., Singh, R. P., Suman, R., & Qadri, M. A. (2023). A pervasive study on green manufacturing towards attaining sustainability. *Green Technologies and Sustainability*, *1*(2), 100018.

Henderson, R. M., & Nellemann, F. (2011). Sustainable tea at Unilever. *Harvard Business School Marketing Unit Case No. 712–438.*

Hetzel, R. (2021). *Giga shed urbanism: Reimagining the Tesla gigafactory as civic space* (Master's thesis, State University of New York at Buffalo).

Hsu, C. C., Choon Tan, K., Hanim Mohamad Zailani, S., & Jayaraman, V. (2013). Supply chain drivers that foster the development of green initiatives in an emerging economy. *International Journal of Operations & Production Management*, *33*(6), 656–688.

Ionescu-Somers, A., & Seifert, R. W. (2014). Leapfrogging to mainstream; Unilever makes lipton tea sustainable. *Supply Chain Forum: An International Journal*, *15*(1), 68–75.

Iqbal, M. W., Kang, Y., & Jeon, H. W. (2020). Zero waste strategy for green supply chain management with minimization of energy consumption. *Journal of Cleaner Production*, *245*, 118827.

Jennings, P. D., & Zandbergen, P. A. (1995). Ecologically sustainable organizations: An institutional approach. *Academy of Management Review*, *20*(4), 1015–1052.

Jo, D., & Kwon, C. (2021). Structure of green supply chain management for sustainability of small and medium enterprises. *Sustainability*, *14*(1), 50.

Kara, S., Hauschild, M., Sutherland, J., & McAloone, T. (2022). Closed-loop systems to circular economy: A pathway to environmental sustainability?. *CIRP Annals*, *71*(2), 505–528.

Karim, R., Roshid, M., & Waaje, A. (2024). Circular economy and supply chain sustainability. In S. S. Phadnis, Y. Sheffi, & C. Caplice (Eds.), *Strategic Innovations for Dynamic Supply Chains* (pp. 1–30). IGI Global.

Khandare, D. M. (2024). Green supply chain management: Strategies for eco-friendly business practices. *Journal of Advanced Zoology*, *45*, 215–223.

Kirchherr, J., Yang, N. H. N., Schulze-Spüntrup, F., Heerink, M. J., & Hartley, K. (2023). Conceptualizing the circular economy (revisited): An analysis of 221 definitions. *Resources, Conservation and Recycling*, *194*, 107001.

Koszewska, M., & Bielecki, M. (2020). How to make furniture industry more circular? The role of component standardisation in ready-to-assemble furniture. *Entrepreneurship and Sustainability Issues*, *7*(3), 1688.

Lerman, L. V., Benitez, G. B., Müller, J. M., de Sousa, P. R., & Frank, A. G. (2022). Smart green supply chain management: A configurational approach to enhance green performance through digital transformation. *Supply Chain Management: An International Journal*, *27*(7), 147–176.

Liu, J., Feng, Y., Zhu, Q., & Sarkis, J. (2018). Green supply chain management and the circular economy: Reviewing theory for advancement of both fields. *International Journal of Physical Distribution & Logistics Management*, *48*(8), 794–817.

Mallick, P. K., Pigosso, D. C., Salling, K. B., & Mcaloone, T. C. (2022). Designing configurator for take-back for a circular economy – A conceptual framework. *Proceedings of the Design Society*, *2*, 1041–1050.

Maradin, D., Malnar, A., & Kaštelan, A. (2022). Sustainable and clean energy: The case of Tesla company. *Journal of Economics, Finance and Management Studies*, *5*(12), 3531–3542.

Miles, S. (2017). Stakeholder theory classification: A theoretical and empirical evaluation of definitions. *Journal of Business Ethics*, *142*, 437–459.

Murphy, P. E., & Murphy, C. E. (2018). Sustainable living: Unilever. In E. O'Higgins, & L. Zsolnai (Eds.), *Progressive Business Models: Creating Sustainable and Pro-social Enterprise* (263–286). Springer.

Oliver, R. K., & Webber, M. D. (1982). Supply chain management: Logistics catches up with strategy. In: M. Christopher (Ed.), *Logistics: The Strategic Issues* (pp. 63–75). Chapman and Hall.

Omair, M., Alkahtani, M., Ayaz, K., Hussain, G., & Buhl, J. (2022). Supply chain modelling of the automobile multi-stage production considering circular economy by waste management using recycling and reworking operations. *Sustainability*, *14*(22), 15428.

Park, S. H. (2020). A case study on the corporate social responsibility in Patagonia 'Worn wear'. *Journal of the Korea Fashion and Costume Design Association*, *22*(1), 61–71.

Patil, S., Jadhav, P., Shinde, D., Pawar, S., Pawar, R., & Chinchanikar, S. (2024). Green manufacturing: A global perspective. In M. K. Dikshit, N. Khanna, A. Soni, & A. P. Markopoulos (Eds.), *International Conference on Futuristic Advancements in Materials, Manufacturing and Thermal Sciences* (pp. 333–350). Springer Nature Singapore.

Pimenov, D. Y., Mia, M., Gupta, M. K., Machado, Á. R., Pintaude, G., Unune, D. R., ... & Kuntoğlu, M. (2022). Resource saving by optimization and machining environments for sustainable manufacturing: A review and future prospects. *Renewable and Sustainable Energy Reviews*, *166*, 112660.

Radhakrishnan, S. (2021). Circular economy in textiles and fashion. In S. S. Muthu (Ed.), *Circular Economy: Assessment and Case Studies* (pp. 163–202). Springer.

Rattalino, F. (2018). Circular advantage anyone? Sustainability-driven innovation and circularity at Patagonia, Inc. *Thunderbird International Business Review*, *60*(5), 747–755.

Risi, D., Vigneau, L., Bohn, S., & Wickert, C. (2023). Institutional theory-based research on corporate social responsibility: Bringing values back in. *International Journal of Management Reviews*, *25*(1), 3–23.

Roy, M. (2025). Harnessing technology and leadership for sustainable agri-food supply chains: Advancing circular economy practices. In D. Burrell, & C. Nguyen (Eds.), *New Horizons in Leadership: Inclusive Explorations in Health, Technology, and Education* (pp. 435–474). IGI Global Scientific Publishing.

Roy, M., & Akbar, D. (2025). Pro-environmental behaviour to minimize food waste in agricultural supply chain. In C. Carrasco-Garrido, A. Martínez Raya, & C. De-Pablos-Heredero (Eds.), *Global Economic Challenges and Opportunities for the Workplace* (pp. 49–88). IGI Global Scientific Publishing.

Roy, M., Akbar, D., Rahman, A., & Rajapaksa, D. (2025a). Trend and pattern of agri-food waste, environmental sustainability and circular economy studies: A science mapping review. *Circular Economy and Sustainability*, 5, 5065–5111.

Roy, M., Akbar, D., & Rajapaksa, D. (2025b). Circular economy principles and their application in agri-food supply chain: A case of Australia. In E. Koç (Ed.), *Developing Dynamic and Sustainable Supply Chains to Achieve Sustainable Development Goals* (pp. 239–282). IGI Global Scientific Publishing.

Roy, M., & Medhekar, A. (2024). Green supply chain, circular economy, and technological innovation: Challenges and opportunities for developing countries. In J. Martínez-Falcó, B. Marco-Lajara, E. Sánchez-García, & L. A. Millán-Tudela (Eds.), *Green Supply Chain Management Practice and Principles* (pp. 102–124). IGI Global Scientific Publishing.

Roy, M., & Sarker, J. (2025). Integration of circular economy principles within information logistics: Case study of Australia. In D. N. Burrell (Ed.), *Organizational Readiness and Research: Security, Management, and Decision Making* (pp. 449–490). IGI Global Scientific Publishing.

Solomon, N. O., Simpa, P., Adenekan, O. A., & Obasi, S. C. (2024). Circular economy principles and their integration into global supply chain strategies. *Finance & Accounting Research Journal*, *6*(5), 747–762.

Srivastava, S. K. (2007). Green supply-chain management: A state-of-the-art literature review. *International Journal of Management Reviews*, *9*(1), 53–80.

Sustainable Manufacturing Report (2025). *Sustainable manufacturing Report 2025: Key Insights & Innovations*. https://www.startus-insights.com/innovators-guide/sustainable-manufacturing-report/ (accessed on 26 April 2026).

Tari, A., & Trudel, R. (2024). Affording disposal control: the effect of circular take-back programs on psychological ownership and valuation. *Journal of Marketing*, *88*(3), 110–126.

United Nations Environment Programme, & International Solid Waste Association (2024). *Global waste Management outlook 2024 – Beyond an Age of Waste: Turning Rubbish into a Resource*. https://wedocs.unep.org/20.500.11822/44939 (accessed on 1 April 2026).

Vachon, S., & Klassen, R. D. (2006). Extending green practices across the supply chain: the impact of upstream and downstream integration. *International Journal of Operations & Production Management*, *26*(7), 795–821.

Vangeri, A. K., Bathrinath, S., Anand, M. C. J., Shanmugathai, M., Meenatchi, N., & Boopathi, S. (2024). Green supply chain management in eco-friendly sustainable manufacturing industries. In J. Arun, N. Nirmala, & S. S. Dawn (Eds.), *Environmental Applications of Carbon-Based Materials* (pp. 253–287). IGI Global.

Wang, Y., Ni, Z., Hu, M., Li, J., Wang, Y., Lu, Z., … & Xia, B. (2020). Environmental performances and energy efficiencies of various urban green infrastructures: A life-cycle assessment. *Journal of Cleaner Production*, *248*, 119244.

Wannags, L. L., & Gold, S. (2020). Assessing tensions in corporate sustainability transition: From a review of the literature towards an actor-oriented management approach. *Journal of Cleaner Production*, *264*, 121662.

Zahran, S. (2024). Investigating the nexus between green supply chain practices and sustainable waste management in advancing circular economy. *Sustainability*, *16*(9), 3566.

Zhang, G., Yang, Y., & Yang, G. (2023). Smart supply chain management in Industry 4.0: The review, research agenda and strategies in North America. *Annals of Operations Research*, *322*(2), 1075–1117.

Zhu, Q., Geng, Y., Fujita, T., & Hashimoto, S. (2010). Green supply chain management in leading manufacturers: Case studies in Japanese large companies. *Management Research Review*, *33*(4), 380–392.

Zucker, L. G. (1987). Institutional theories of organization. *Annual Review of Sociology*, *13*, 443–464.

9 A Hybrid Systematic Literature Review on RL in Furniture Industry

Saleh Md. Arman and Md. Mamun Habib

9.1 INTRODUCTION

Reverse logistics (RL) addresses these viable issues by enabling the recovery, refurbishment, and recycling of materials, which minimize landfill waste and conserves valuable resources. RL is the strategic approach of retrieving, processing, and reintroducing products into the supply chain, or disposing of them responsibly to minimize environmental impact (Susanty et al., 2020). It merges economic and ecological priorities by promoting product returns, refurbishments, and end-of-life management aligned with sustainable development goals (He et al., 2024). According to Garzon Agudelo et al. (2021; p. 3), RL in the furniture industry is the process that "… begins with the physical return of materials, which may be classified according to the possibility of recycling and finally determine the criteria for use or reuse, as appropriate."

RL is gaining popularity in many industries (Iakovou et al., 2024). Industries rely on it to improve operational efficiency and strengthen customer retention (Butt et al., 2024). This multidisciplinary area addresses sustainability issues, uncovers new research frontiers, and initiates industry collaborations (Arman & Mark-Herbert, 2021). However, scholarly knowledge in the furniture sector remains fragmented, lacking a synthesized interdisciplinary perspective that advances the research field and answers the questions raised by industry practitioners. A systematic literature review integrates this dispersed knowledge, laying a strong foundation for strategic improvements, industry innovation, and academic progress (Paul et al., 2021).

The objective of the book chapter is to conduct the systematic literature review to establish a robust foundation for strategic improvements, industry innovation, and future academic research by uncovering critical research gaps. Based on this research objective, the study seeks to address the following research questions.

RQ1. *What are the antecedents, mediators, and moderators influencing RL in the furniture industry?*

RQ2. *In what ways do academic theories, research contexts, and methods impact RL in the furniture industry?*

RQ3. *What are the research gaps in the existing literature on RL in the furniture industry?*

DOI: 10.1201/9781003770008-13

RQ4. *What are the future research propositions on RL in the furniture industry?*

The research novelty is manifold. This systematic literature review on RL in the furniture industry consolidates insights into key research areas. By synthesizing diverse findings, it establishes a comprehensive knowledge base that guides stakeholders toward innovative, resource-efficient solutions. Also, the study identifies research gaps for directing further research in order to strengthen the research field.

The rest of the sections of this book chapter is organized as follows. Section 9.2 describes the theoretical background. Section 9.3 identifies the research methodology, Section 9.4 presents result from the analysis and discussions, Section 9.5 presents the future research directions on the basis of research gaps. The last section of the book chapter presents conclusions.

9.2 THEORETICAL BACKGROUND

RL focuses on efficiently moving products from consumers back to the producers, or to appropriate disposal facilities for minimizing waste (Dhanorkara, 2019; Xie et al., 2025). Product manufacturers adopt it to manage returns, refurbish products, and recycle materials (Başaran, 2013; Lin et al., 2020). This process enhances supply chain sustainability and reduces waste through strategic reclamation of value from end-of-life products (Accorsi et al., 2015; Soleimani & Kannan, 2015). By yielding environmental and economic benefits, it supports companies in fulfilling corporate social responsibilities (Björklund et al., 2024).

In the furniture industry, RL addresses high return rates due to size and style mismatches (Sellitto et al., 2017). By reclaiming, repairing, and reselling products, they tap into secondary markets that promote customers' ethical consumption behavior and reduce the furniture sector's dependence on virgin raw materials (Arman & Mark-Herbert, 2021; Dhanorkara, 2019). Besides environment-friendly functions of RL, technological enhancement made this process more effective. This smooth experience positively reinforces the company's brand image and customer loyalty (Starostka-Patyk, 2019).

9.3 METHODOLOGY

This study employs the Antecedent Decision Outcome – Theory Context Method, i.e., ADO-TCM framework by Lim et al. (2021). ADO-TCM serves as an integrative approach that broadens and deepens reviews, mitigating the limitations inherent to each constituent framework (Arman & Mark-Herbert, 2024). This framework was previously used in different SCM applications notably green logistics practices (Tetteh et al., 2024), power in the supply chains (Zaher & Marquez-Illescas, 2024), and big data analytics in SCM decarbonization (Kumar et al., 2024). The research utilizes the PRISMA 2020 protocol for performing the systematic literature review (Page et al., 2021). In the initial phase of the research process, the researchers chose Scopus and Web of Science databases (Arman & Mark-Herbert, 2024). The search query utilized the keywords related to RL and furniture (see Figure 9.1). The timeline for the selected documents is 2010–July 23, 2025. The study outlines the protocol in Figure 9.1.

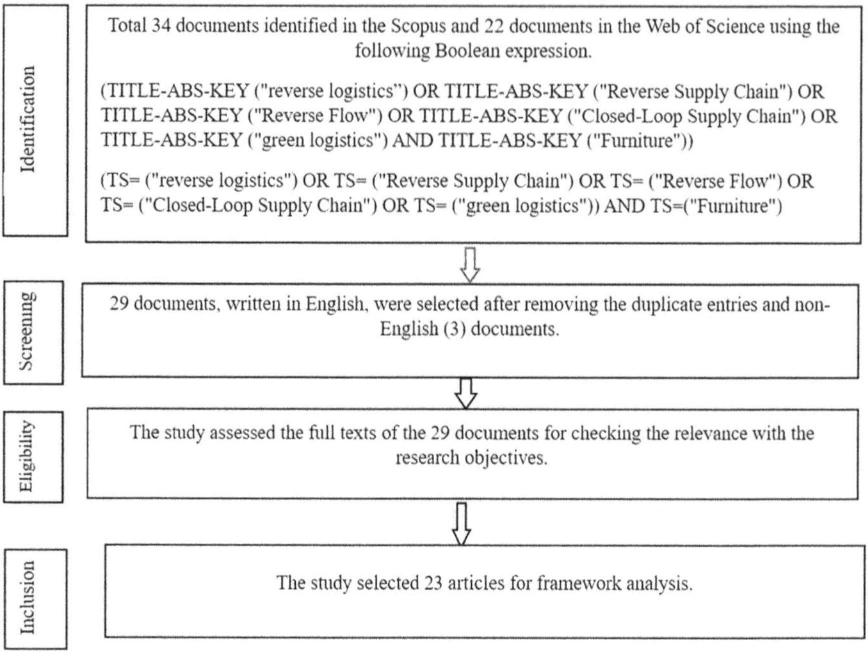

FIGURE 9.1 PRISMA 2020 Protocol for systematic literature review.

The study read the full texts of the finally selected 23 papers to find out the underlying antecedents, mediators, moderators, barriers for constructing the conceptual framework. Also, the study identified theories, research contexts, and research methods from the full texts to find out the research gaps that lead to design future research directions.

9.4 DISCUSSION

9.4.1 Findings from the ADO Framework – RQ1

Consumer awareness is the most frequently used antecedents which is found in 11 documents. Consumers are increasingly aware of environmental impacts and actively support closed-loop supply chains that repurpose or recycle used furniture (Björklund et al., 2024; Garzon Agudelo et al., 2021; Putthinoi et al., 2015). As consumer demand for eco-friendly solutions rises, companies are motivated to invest in RL systems that reduce waste and optimize resource use (Soleimani & Kannan, 2015; Susanty et al., 2020; Zhang et al., 2022). **Policies designed by the regulatory agencies** are discussed in nine documents. Regulatory agencies set standards for the materials' safe-handling, storage, and disposal of furniture waste, thereby influencing RL adoption (dos Santos et al., 2019; Garzon Agudelo et al., 2021; Zhang et al., 2022). Other antecedents are **environmental legislation** (Accorsi et al., 2015; Erol et al., 2010; Sepúlveda-Rojas et al., 2024; Soleimani & Kannan, 2015; Susanty et al.,

2020; Zhang et al., 2022), **economic factors** (Accorsi et al., 2015; Dhanorkara, 2019; Ocampo et al., 2019; Sagnak, 2021; Soleimani & Kannan, 2015), **demographic factors** (Dhanorkara, 2019; Ocampo et al., 2019; Putthinoi et al., 2015; Sellitto et al., 2017), **geographical factors** (Accorsi et al., 2015; Björklund et al., 2024; Dhanorkara, 2019), **advancements in technology** (Başaran, 2013; Björklund et al., 2024; Dhanorkara, 2019; Franco, 2019; Koszewska & Bielecki, 2020; Prakash et al., 2017), **business process models** (Erol et al., 2010; Franco, 2019; Koszewska & Bielecki, 2020; Krystofik et al., 2018; Sagnak, 2021), and **furniture industry-specific components** (Başaran, 2013; Garzon Agudelo et al., 2021; Koszewska & Bielecki, 2020; Krystofik et al., 2018; Ocampo et al., 2019; Ramanathan & Ramakrishnan, 2020; Sellitto et al., 2017).

9.4.1.1 Decisions

As a mediator, **consumer perception** influences the strategic decisions companies make regarding resource recovery and waste reduction (Björklund et al., 2024; Garzon Agudelo et al., 2021; Putthinoi et al., 2015). Additionally, **logistics service providers** (Björklund et al., 2024; dos Santos et al., 2019; Prakash et al., 2017; Ramanathan & Ramakrishnan, 2020), **regulatory agencies** (Koszewska & Bielecki, 2020; Ocampo et al., 2019; Ramanathan & Ramakrishnan, 2020), and **stakeholder involvement** (Başaran, 2013; Björklund et al., 2024; dos Santos et al., 2019) mediates the relationships toward achieving sustainable development goals. As a moderator, **regulatory agencies** set standards and requirements that guide empirical inquiry, promoting transparency, and fostering innovation in RL solutions. Also, **consumer awareness** (Minnema et al., 2016), and **competitors Awareness** (Erol et al., 2010) positively influence RL decisions. The existing pressures from consumers and regulatory agencies are for environmentally friendly products. Lastly, the barriers are **system inadequacies** that hinder the needed shifts in the business practices (Erol et al., 2010), **lack of knowledge and awareness** of RL strategies of stakeholders (Garzon Agudelo et al., 2021; Prakash et al., 2017), and **insufficient green emphasis** in existing logistics activities, which collectively limit the RL adoption and integration within the furniture industry (Sagnak, 2021).

9.4.1.2 Outcome

Choosing the right design and business model shapes consumer behavior and end-of-life product management (Franco, 2019). The RL management decision in furniture industry from leaders to laggards, where their engagement level with environmentally oriented supply chain cooperation (ESCC) dictates how effectively they adopt circular economy (CE) practices (Susanty et al., 2020). Finally, companies that integrate environmental management systems demonstrate managerial commitment and companies that adopt credible certifications often realize stronger market competitiveness and resource efficiencies.

9.4.1.3 A Conceptual Framework

Based on the discussion, the study designs the following conceptual framework for RL applications in the furniture industry (Figure 9.2).

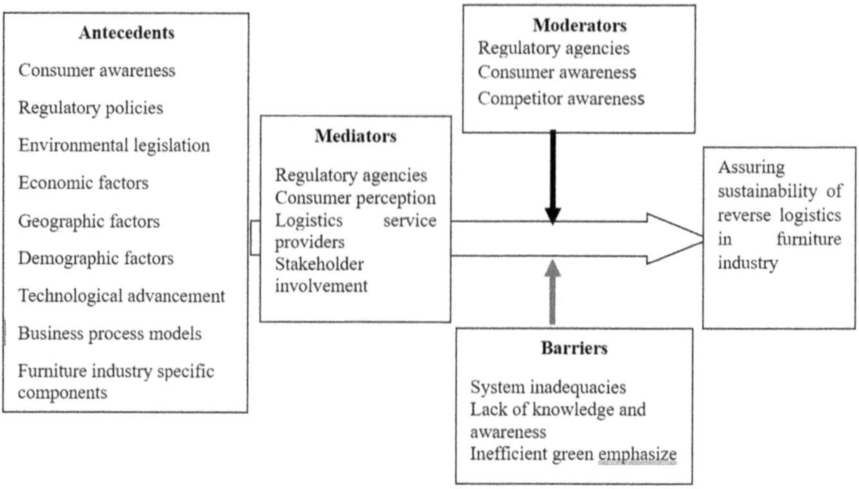

FIGURE 9.2 A conceptual framework of RL operations in furniture industry.

9.4.2 Findings from the TCM Framework – RQ2

9.4.2.1 Theories

Most of the authors used conceptual framework in terms of RL operations in furniture industry. Soleimani & Kannan (2015) used a hybrid PSO-GA (Particle Swarm Optimization-Genetic Algorithm), Minnema et al. (2016) designed conceptual framework in the combination of McFadden (1974)'s utility theory, Rust et al. (1999)'s customer expectations and uncertainty, Oliver (2010)'s satisfaction perspectives and Zeithaml et al. (1993)'s customer expectations formation and named it as "expectation disconfirmation theory."

However, the extended Theory of Planned Behavior (Başaran, 2013), The Ellen MacArthur Foundation's (2013) circular economy theory (Koszewska & Bielecki, 2020), stakeholder theory (Ramanathan & Ramakrishnan, 2020), and risk management theory (Prakash et al., 2017) were also used. These theories collectively reveal that effective RL in furniture industry requires integrating consumer behavior, circular practices, stakeholder collaboration, and risk mitigation to enhance sustainability outcomes. Such integrations can reduce waste, improve resource recovery, and boost economic viability, though implementation varies by region and company size.

9.4.2.2 Contexts

The country contextual coverage of the research on RL in furniture industry is shown in Figure 9.3.

From the figure, it showed that the selected studies were concentrated on the following countries: Brazil, Germany, Japan, China (2), Indonesia, Columbia, Chile, Turkey (3), UK, Sweden, Poland, Thailand, Taiwan, India, USA (4), Brazil, Philippines, Italy, Iran. It means that the RL research contexts are not limited to any

FIGURE 9.3 Research contexts.

TABLE 9.1
Research Methods

Methods	References
Quantitative	Accorsi et al. (2015), Başaran (2013), Franco (2019), Koszewska & Bielecki (2020), Lin et al. (2020), Minnema et al. (2016), Prakash et al. (2017), Sepúlveda-Rojas et al. (2024), Soleimani & Kannan (2015), Susanty et al. (2020)
Qualitative	Björklund et al. (2024), Putthinoi et al. (2015), Ramanathan & Ramakrishnan (2020), Sagnak (2021), Sellitto et al. (2017)
Mixed	Erol et al. (2010), dos Santos et al. (2019), Garzon Agudelo et al. (2021), Krystofik et al. (2018), Ocampo et al. (2019), Starostka-Patyk (2019), Zhang et al. (2022)

fix continents. Rather this is an universal research contexts. Most researchers concentrated on specific countries, but Koszewska & Bielecki (2020) conducted study on EU countries, but they didn't specify any county names.

9.4.2.3 Methods

Most of the studies use quantitative methods in conducting research. Table 9.1 mentions the methods used in the research on RL in furniture industry.

9.5 FUTURE RESEARCH DIRECTIONS – RQ3 AND RQ4

The study assesses research gaps according to necessity, importance, relevance, and urgency, guided by Miles's (2017) seven-gap typology: population, evidence, knowledge, practical-knowledge, methodology, empirical, and theoretical.

9.5.1 Research Questions Based on ADO Framework

The proposed framework presents several antecedents of RL in furniture industry that are crucial for assuring sustainability in the supply chain. However, the product types are not fully explored in furniture industry perspectives (Accorsi et al., 2015; Ocampo et al., 2019). Future research could investigate how different types of products correspond differently in terms of product characteristics offered by the furniture industry. There is also potential research gap for examining long-term effects on consumer behavior beyond immediate purchase and furniture return decisions (Accorsi et al., 2015). Moreover, Erol et al. (2010) pointed out the insufficient exploration of consumer attitudes toward RL and their willingness to participate in product recovery and remanufacturing initiatives. Başaran (2013) and Dhanorkara (2019) points on exploring additional variables that may influence recycling behaviors, for example, cultural factors influencing consumer decision-making.

RQ1. *How do product characteristics of different furniture types influence the effectiveness of RL strategies?*
RQ2. *How do cultural factors shape consumer attitudes in the RL?*

Additionally, this study points at gaps which could address the significance of the mediators and moderators of the RL decision-making and prevent the barriers. Minnema et al. (2016) and Accorsi et al. (2015) point out the gap of focusing on the influence on product return decisions by end-consumers. To address these gaps, the study proposes the following research questions.

RQ3. *In what ways do product quality and pricing interact with online consumer review characteristics to influence return behavior among furniture buyers?*

For assuring sustainable outcomes, Sepúlveda-Rojas et al. (2024) and Garzon Agudelo et al. (2021) point on the lack of sustainability alignment in corporate strategies that hinders the desired outcome. Based on the discussion, the study designs the last research questions.

RQ4. *How does the sustainability alignment influence supplier selection?*

9.5.2 Research Objectives Based on TCM Framework

By following the TCM framework, the study addresses the gaps in terms of theories used, research contexts, and research methods. The study identifies the call for designing a comprehensive theory dedicated to RL in furniture industry to address the uncertainty, sustainability, and stakeholder selection and relationships (Table 9.2) (Dhanorkara, 2019; Erol et al., 2010; Franco, 2019; Koszewska & Bielecki, 2020; Prakash et al., 2017; Sagnak, 2021; Starostka-Patyk, 2019). Based on the discussion, the study proposes the following two research objectives.

TABLE 9.2
Future Research Directions Using Gap Analysis and ADO-TCM Framework

	Research Gap	Gap Type	Future Research Questions
Where Should We Be Heading			
Antecedents	Characteristics of products not explored Consumer recycling behaviors and its long-term effects	Empirical Practical-knowledge	**RQ1**. How do product characteristics influence the effectiveness of RL strategies? **RQ2**. How do cultural factors shape consumer attitudes in the RL?
Decisions	Lack of research on product return decisions	Empirical	**RQ3**. In what ways do product quality and pricing influence return behavior among furniture buyers?
Outcome	Need sustainability alignment in RL supply chains	Empirical evidence	**RQ4**. How does the sustainability alignment influence supplier selection?
How Can We Get There			
Theories	Need for integrated economic, environmental, and social sustainability research.	Knowledge method	**RO1**. Develop an integrated theoretical framework for the furniture industry.
Contexts	Consumer demographics are not fully explored Cross-country research is inadequate	Empirical Knowledge	**RO2**. Investigate the impact of consumer demographics in the furniture industry. **RO3**. Conduct comparative cross-country research to explore the different RL practices in the furniture industry.
Methods	Insufficient methodologies with practical applicability.	Method	**RO4**. Design and validate a practical, risk-integrated optimization model for RL in the furniture industry.

RO1. *Develop an integrated theoretical framework that incorporates economic, environmental, and social sustainability aspects into RL for the furniture industry.*

In addition, present studies identify that consumer demographics are essential in the research context but remain underexplored as there should be some discrepancies in terms of age, gender, race, education, and professional background (Accorsi et al., 2015; Putthinoi et al., 2015; Ramanathan & Ramakrishnan, 2020). Also, several studies also point out the gap of the lack of empirical studies in developing country contexts (Başaran, 2013; dos Santos et al., 2019; Franco, 2019; Garzon Agudelo et al., 2021; Ocampo et al., 2019). Based on the discussion on research contexts, the study proposes the following future research objectives.

RO2. *Investigate the impact of consumer demographics on RL practices.*
RO3. *Conduct comparative cross-country research to explore the different RL practices in the furniture industry.*

Despite diverse research methods, preferably quantitative, deployed by the researchers, limited studies have focused on integrating multiple types of risks that address both forward and RL under uncertainty (Prakash et al., 2017). Based on the discussion on research contexts, the study proposes the following future research objectives.

RO4. *Design and validate a practical, risk-integrated optimization model for RL in the furniture industry.*

9.6 CONCLUSIONS

9.6.1 IMPLICATIONS

This study significantly enhances the theoretical landscape of RL by integrating diverse perspectives and frameworks, particularly the ADO-TCM framework. Building on the fragmented findings of theoretical approaches (Section 9.4.2.1), where most studies relied on conceptual frameworks rather than established theories, the research emphasizes the necessity for a comprehensive theory that encompasses economic, environmental, and social sustainability aspects within the furniture industry. This directly addresses the identified gap (RO1 and RO2) for integrated theoretical frameworks that can guide future research. By identifying key antecedents and barriers, the research establishes a foundation for future theoretical explorations and encourages scholars to investigate the interplay between consumer behavior, product characteristics, and RL strategies in the furniture industry. Researchers can test the proposed conceptual framework through surveys, case studies, or experiments measuring the strength of connections between framework elements.

From a practical standpoint, the findings underscore the importance of aligning corporate sustainability strategies with RL practices. The study revealed that consumer awareness emerged as the most frequent antecedent (11 documents), while system inadequacies and lack of knowledge were major barriers (Section 9.4.1.3), directly informing the practical need for education and system improvements. Additionally, addressing the identified barriers facilitates smoother implementation of RL systems, ultimately leading to improved resource management and waste reduction. Furniture companies can avail the proposed conceptual framework to assess their current RL status by evaluating presence/absence of antecedents, identifying existing barriers and underlying mediators and moderators to strengthen company performance.

This research has profound social implications, as it emphasizes the role of consumer awareness and engagement in promoting sustainable practices. The research findings show significant geographical variations in RL implementation (Figure 9.3), with concentration in certain countries, highlighting the need for cross-country research to ensure equitable access to sustainable furniture practices globally. This directly responds to the identified gap regarding unexplored consumer demographics and their influence on furniture companies. The framework's emphasis on stakeholder involvement as a mediator provides a pathway for translating research into community-based sustainable practices.

9.6.2 Limitations and Future Research Directions

Despite its contributions, this study has limitations, including the reliance on existing literature, which may not capture emerging trends in RL. The study focuses on RL operations furniture industry only which itself is a niche area. Hence, there is a limited number of empirical studies that this systematic literature review considered. Future research should consider diversified empirical studies that explore the long-term effects of RL practices on consumer behavior and furniture company performance. Additionally, investigating the impact of digitalization and online platforms on RL can provide valuable insights. Expanding research to include diverse geographical contexts, particularly in developing countries, will enhance the understanding of global RL practices and their implications for sustainability in the furniture industry.

REFERENCES

Accorsi, R., Manzini, R., Pini, C., & Penazzi, S. (2015). On the design of closed-loop networks for product life cycle management: Economic, environmental and geography considerations. *Journal of Transport Geography*, *48*, 121–134. https://doi.org/10.1016/j.jtrangeo.2015.09.005

Arman, S. M., & Mark-Herbert, C. (2021). Re-commerce to ensure circular economy from consumer perspective. *Sustainability*, *13*(18), 18. https://doi.org/10.3390/su131810242

Arman, S. M., & Mark-Herbert, C. (2024). Ethical consumption: A review and research agenda. *International Journal of Consumer Studies*, *48*(5), e13079. https://doi.org/10.1111/ijcs.13079

Başaran, B. (2013). What makes manufacturing companies more desirous of recycling? *Management of Environmental Quality: An International Journal*, *24*(1), 107–122. https://doi.org/10.1108/14777831311291177

Björklund, M., Gustafsson, S., & Skill, K. (2024). Sustainability potentials of digitally based platforms for the circularity of household items. *Cleaner Logistics and Supply Chain*, *10*, 100133. https://doi.org/10.1016/j.clscn.2023.100133

Butt, A. S., Ali, I., & Govindan, K. (2024). The role of reverse logistics in a circular economy for achieving sustainable development goals: A multiple case study of retail firms. *Production Planning & Control*, *35*(12), 1490–1502. https://doi.org/10.1080/09537287.2023.2197851

Dhanorkara, S. (2019). Environmental benefits of internet-enabled C2C closed-loop supply chains: A quasi-experimental study of craigslist. *Management Science*, *65*, 660–680. https://doi.org/10.1287/mnsc.2017.2963

dos Santos, B. M., Godoy, L. P., & Campos, L. M. S. (2019). Performance evaluation of green suppliers using entropy-TOPSIS-F. *Journal of Cleaner Production*, *207*, 498–509. https://doi.org/10.1016/j.jclepro.2018.09.235

Erol, İ., Nurtaniş Velioğlu, M., Sivrikaya Şerifoğlu, F., Büyüközkan, G., Aras, N., Demircan Çakar, N., & Korugan, A. (2010). Exploring reverse supply chain management practices in Turkey. *Supply Chain Management: An International Journal*, *15*(1), 43–54. https://doi.org/10.1108/13598541011018111

Franco, M. A. (2019). A system dynamics approach to product design and business model strategies for the circular economy. *Journal of Cleaner Production*, *241*, 118327. https://doi.org/10.1016/j.jclepro.2019.118327

Garzon Agudelo, P., Palacios-Alvarado, W., & Medina Delgado, B. (2021). Diagnosis of physical conditions for the implementation of a reverse logistics management model in a supply chain. *Journal of Physics: Conference Series*, *1938*, 012019. https://doi.org/10.1088/1742-6596/1938/1/012019

He, M., Li, Q., Wu, X., & Han, X. (2024). A novel multi-level reverse logistics network design optimization model for waste batteries considering facility technology types. *Journal of Cleaner Production*, *467*, 142966. https://doi.org/10.1016/j.jclepro.2024.142966

Iakovou, E., Pistikopoulos, E. N., Walzberg, J., Iseri, F., Iseri, H., Chrisandina, N. J., Vedant, S., & Nkoutche, C. (2024). Next-generation reverse logistics networks of photovoltaic recycling: Perspectives and challenges. *Solar Energy*, *271*, 112329. https://doi.org/10.1016/j.solener.2024.112329

Koszewska, M., & Bielecki, M. (2020). How to make furniture industry more circular? The role of component standardisation in ready-to-assemble furniture. *Entrepreneurship and Sustainability Issues*, *7*(3), 1688–1707. https://doi.org/10.9770/jesi.2020.7.3(17)

Krystofik, M., Luccitti, A., Parnell, K., & Thurston, M. (2018). Adaptive remanufacturing for multiple lifecycles: A case study in office furniture. *Resources, Conservation and Recycling*, *135*, 14–23. https://doi.org/10.1016/j.resconrec.2017.07.028

Kumar, D., Singh, R. K., Mishra, R., & Vlachos, I. (2024). Big data analytics in supply chain decarbonisation: A systematic literature review and future research directions. *International Journal of Production Research*, *62*(4), 1489–1509. https://doi.org/10.1080/00207543.2023.2179346

Lim, W. M., Yap, S.-F., & Makkar, M. (2021). Home sharing in marketing and tourism at a tipping point: What do we know, how do we know, and where should we be heading? *Journal of Business Research*, *122*, 534–566. https://doi.org/10.1016/j.jbusres.2020.08.051

Lin, C.-W. R., Chen, M.-T., Tseng, M.-L., Chiu, A. S. F., & Ali, M. H. (2020). Profit maximization for waste furniture recycled in Taiwan using cradle-to-cradle production programming. *Mathematical Problems in Engineering*, *2020*(1), 2948049. https://doi.org/10.1155/2020/2948049

McFadden, D. (1974). The measurement of urban travel demand. *Journal of Public Economics*, *3*(4), 303–328. https://doi.org/10.1016/0047-2727(74)90003-6

Miles, D. A. (2017). A taxonomy of research gaps: identifying and defining the seven research gaps. *Journal of Research Methods and Strategies*. https://www.academia.edu/35505149/ARTICLE_RESEARCH_A_Taxonomy_of_Research_Gaps_Identifying_and_Defining_the_Seven_Research_Gaps

Minnema, A., Bijmolt, T. H. A., Gensler, S., & Wiesel, T. (2016). To keep or not to keep: Effects of Online customer reviews on product returns. *Journal of Retailing*, *92*(3), 253–267. https://doi.org/10.1016/j.jretai.2016.03.001

Ocampo, L., Himang, C., Kumar, A., & Brezocnik, M. (2019). A novel multiple criteria decision-making approach based on fuzzy DEMATEL, fuzzy ANP and fuzzy AHP for mapping collection and distribution centers in reverse logistics. *Advances in Production Engineering & Management*, *14*, 297–322. https://doi.org/10.14743/apem2019.3.329

Oliver, R. L. (2010). *Satisfaction: A Behavioral Perspective on the Consumer*. M.E. Sharpe.

Page, M. J., McKenzie, J. E., Bossuyt, P. M., Boutron, I., Hoffmann, T. C., Mulrow, C. D., Shamseer, L., Tetzlaff, J. M., & Moher, D. (2021). Updating guidance for reporting systematic reviews: Development of the PRISMA 2020 statement. *Journal of Clinical Epidemiology*, *134*, 103–112. https://doi.org/10.1016/j.jclinepi.2021.02.003

Paul, J., Lim, W. M., O'Cass, A., Hao, A. W., & Bresciani, S. (2021). Scientific procedures and rationales for systematic literature reviews (SPAR-4-SLR). *International Journal of Consumer Studies*, *45*(4), O1–O16. https://doi.org/10.1111/ijcs.12695

Prakash, S., Soni, G., & Rathore, A. P. S. (2017). Embedding risk in closed-loop supply chain network design: Case of a hospital furniture manufacturer. *Journal of Modelling in Management*, *12*(3), 551–574. https://doi.org/10.1108/JM2-02-2016-0017

Putthinoi, S., Sachon, S., Yodmongkol, P., & Chakpitak, N. (2015). Household reverse logistics analysis using the SCOR model to improve home safety of the elderly in the municipality. *International Journal of Logistics Systems and Management*, *21*, 348. https://doi.org/10.1504/IJLSM.2015.069732

Ramanathan, U., & Ramakrishnan, R. (2020). *Sustainable Supply Chains: Strategies, Issues, and Models*. Springer. https://doi.org/10.1007/978-3-030-48876-5

Rust, R. T., Inman, J. J., Jia, J., & Zahorik, A. (1999). What you don't know about customer-perceived quality: The role of customer expectation distributions. *Marketing Science*, *18*(1), 77–92. https://doi.org/10.1287/mksc.18.1.77

Sagnak, M. (2021). Assessment of logistics performance in sustainable supply chain: Case from emerging economy. In S. K. Mangla, & M. Ram (Eds.), *Supply Chain Sustainability: Modeling and Innovative Research Frameworks* (pp. 73–88). Walter De Gruyter. https://doi.org/10.1515/9783110628593-004

Sellitto, M., Luchese, J., Bauer, J., Saueressig, G., & Viegas, C. (2017). Ecodesign practices in a furniture industrial cluster of southern Brazil: From incipient practices to improvement. *Journal of Environmental Assessment Policy and Management*, *19*, 1750001. https://doi.org/10.1142/S1464333217500016

Sepúlveda-Rojas, J. P., Aravena, S., & Carrasco, R. (2024). Increasing efficiency in furniture remanufacturing with AHP and the SECI model. *Sustainability*, *16*(23), 23. https://doi.org/10.3390/su162310339

Soleimani, H., & Kannan, G. (2015). A hybrid particle swarm optimization and genetic algorithm for closed-loop supply chain network design in large-scale networks. *Applied Mathematical Modelling*, *39*(14), 3990–4012. https://doi.org/10.1016/j.apm.2014.12.016

Starostka-Patyk, M. (2019). Defective products management with reverse logistics processes in the furniture production companies. *Polish Journal of Management Studies*. https://doi.org/10.17512/pjms.2019.20.2.42

Susanty, A., Tjahjono, B., & Sulistyani, R. E. (2020). An investigation into circular economy practices in the traditional wooden furniture industry. *Production Planning & Control*, *31*(16), 1336–1348. https://doi.org/10.1080/09537287.2019.1707322

Tetteh, F. K., Owusu Kwateng, K., & Mensah, J. (2024). Green logistics practices: A bibliometric and systematic methodological review and future research opportunities. *Journal of Cleaner Production*, *476*, 143735. https://doi.org/10.1016/j.jclepro.2024.143735

Xie, Y., Arman, S. M., & Su, C. (2025). Circular economy in post consumption network: The role of re-commerce groups in social media platforms. *Electronic Commerce Research and Applications*, *70*, 101489. https://doi.org/10.1016/j.elerap.2025.101489

Zaher, H. F., & Marquez-Illescas, G. (2024). Power in the supply chain: A state-of-the-art literature review and propositions from the perspective of gender differences. *Journal of Business & Industrial Marketing*, *39*(6), 1282–1310. (world). https://doi.org/10.1108/JBIM-10-2022-0484

Zeithaml, V. A., Berry, L. L., & Parasuraman, A. (1993). The Nature and Determinants of Customer Expectations of Service. *Journal of the Academy of Marketing Science*, *21*(1), 1–12. https://doi.org/10.1177/0092070393211001

Zhang, X., Zou, B., Feng, Z., Wang, Y., & Yan, W. (2022). A review on remanufacturing reverse logistics network design and model optimization. *Processes*, *10*(1), 1. https://doi.org/10.3390/pr10010084

Section V

Agricultural and Ecosystem-Based Value Chains

10 Monetizing Ecosystem Services in Supply Chains in Malaysia

The Case for Wetland Conservation

Suziana Hassan, Banji Rildwan Olaleye, and Rauda Md. Ramli

10.1 INTRODUCTION

Wetlands support ecological functions and economic activities but face degradation from industrialization, urbanization, and mismanagement (Salmi et al., 2023). Sustainable supply chain management (SSCM) aims to incorporate environmental protection into business; yet, wetlands remain underintegrated into sustainability models (Mugoni et al., 2024; Olaleye & Mosleh, 2025). This hinders conservation and affects ecosystems-dependent industries. The Setiu Wetlands (SW) face land-use conflicts, pollution, and resource overexploitation, highlighting the need for sustainable management (Lola et al., 2016). While government agencies and NGOs promote conservation, funding remains uncertain. SSCM strategies, like circular economy practices promoting ecotourism, waste reduction, and sustainable fishing, offer practical solutions to balance economic growth and conservation. Green supply chains, including eco-purchasing and carbon-neutral transportation, support sustainability for businesses near wetlands. Although Payment for Ecosystem Services (PES) mechanisms are proposed for conservation funding, there's limited empirical data on their impact on public perception, willingness to pay (WTP), and legitimacy within supply chains.

This chapter explores the combination of payment structure with supply chain management, providing important information for conservation organizations to design strategic financial mechanism for a sustainable management. In achieving this research aim, the five research questions have been formulated, i.e., (a) Which wetland conservation attributes are most valued by the public within an SSCM context?; (b) How does WTP vary across different conservation financing mechanisms?; (c) Are voluntary payment mechanisms more acceptable than coercive instruments?; (d) How do public preferences challenge conventional supply chain risk assumptions? (d) How

DOI: 10.1201/9781003770008-15

can wetland valuation be operationalized within SSCM frameworks? Alternatively, the research objectives are to (a) examine public preferences for key wetland conservation attributes relevant to SSCM; (b) estimate WTP for wetland conservation under different financing mechanisms; (c) compare the financial mechanism such as donation, taxation, and subsidy reduction mechanisms; (d) assess how conservation attributes influence perceived supply chain resilience; and provide empirical guidance for integrating wetland conservation into SSCM strategies.

The results will support other wetland ecosystems facing conservation and economic debates. Empirical research is needed to assess how conservation features and financing strategies together influence public preferences and funding, especially in SSCM frameworks in developing economies. This chapter includes Introduction, Literature Review, Methodology, Results and Discussion, and Conclusion, ending with study limitations and future research opportunities.

10.2 LITERATURE REVIEW

10.2.1 Background

Using the economic valuation approach, governments and businesses could get a better understanding of the importance of preserving wetlands and the effects it has in supply chains over a long period (Dingha et al., 2025). According to Dharmayanti et al. (2023), SSCM also emphasizes that companies need to protect the environment and at the same time, they must maintain business sustainability. This literature review analyses ways in which ecosystem service valuation (ESV) can guide SSCM to the industries that are dependent on wetlands. ESV concept will allow policymakers and stakeholders in the industry to consider the natural environment in their investment decision and long-term planning. People, Planet, Profit are the Triple Bottom Line principles that allow creating a balance of economic profitability with the environmental and social management of the state (Tzanidis et al., 2024). Furthermore, the growth of the circular economy along with green supply chain management will help to produce conservation funding mechanisms that would reflect purposes of supply chain stability, corporate sustainability, and community benefits. On the other hand, PES is a financial tool that associated business motives and conservation goals (Hejnowicz et al., 2014; Le et al., 2024). The works of Chen et al. (2014), Cranford and Mourato (2014), and Hein et al. (2013) prove that PES systems can be effective to draw financial resources to conservation, but its effectiveness is conditional upon the acceptance of the technology and price insensitivity of people.

10.2.2 Problem Statement

Wetlands are crucial for both the health of ecosystems and the economy, but less emphases in making decisions about their supply chains. There are numerous SSCM studies devoted to emission, green logistics, and efficiency of materials. Decreased focus on the significance of ecosystem interconnectedness to ensure industrial productivity. (Gupta and Palsule-Desai, 2011; Shekarian et al., 2022; Tchonkouang et al., 2024; VanderWilde et al., 2023). People generally agree that protecting the environment is good for public health; yet, it is still not properly incorporated into supply chain systems. A recent study in Malaysia on the economic assessment of biodiversity conservation in a national park demonstrates that conservation has a significant impact on social and economic advantages that surpass ecological results (Ibrahim et al., 2025). The absence of empirical research comparing these mechanisms in wetland management generates ambiguity for policymakers and businesses to integrate conservation into SSCM plans (Mugoni et al., 2024). Thus, wetland valuation should be carefully assessed in SSCM to improve ecological resilience and supply chain sustainability.

10.2.3 Scope of the Study

This chapter provides an economic assessment of wetland ecosystem services, focusing SW in Malaysia as a representative of the wetland supply chain system. The focus is narrowed to attributes that can be associated with supply chain management of wetland conservation such as protecting biodiversity, flood risk reduction, buffer zones, and recreational services. This chapter also explores three payment mechanisms, i.e., voluntary donations, taxation, and subsidy reductions, to illustrate how various financial strategies affect public WTP for wetland conservation.

10.2.4 Theoretical Contributions

This study advances theory in three ways. First, it integrates wetland conservation attributes, i.e., biodiversity, flood risk reduction, buffer zones, and recreational services into an SSCM framework by integrating ESV as a strategic supply chain input. Second, it examines WTP across different payment methods to enhance conservation financing. Third, it combines the environmental economics with supply chain elements, which makes SSCM theory more compelling. From these theoretical foundations, we developed the following hypotheses: H1: Conservation characteristics (biodiversity protection, flood mitigation, buffer zones, and recreational facilities) strongly affect public preferences for wetland conservation within SSCM frameworks; H2: The WTP for wetland conservation depends on financial mechanism, for example, donations and taxes are better than reduction of subsidies. H3: People see subsidy reduction as coercive, which makes them less likely to accept them and more sensitive to costs. These hypotheses will be evaluated using empirical data analysis employing a CE method.

10.2.5 Critical Analysis and Research Gaps

SSCM studies like Dharmayanti et al. (2023), Tchonkouang et al. (2024), and VanderWilde et al. (2023) prioritize emissions, efficiency, and circularity but ignore environmental dependencies that pose strategic risks to fisheries and tourism, creating supply chain risk assessment blind spots. Conservation finance options including donations, taxes, and subsidy reductions are rarely compared empirically, especially in developing countries where public approval affects success and equity. The study uses SW in Malaysia to examine public WTP for wetland conservation under three funding mechanisms, i.e., donation, taxation, and subsidy reduction, to fill these gaps. It provides a model for linking conservation fund to sustainable business operations.

10.3 METHODOLOGY

This study employs a quantitative, cross-sectional, and explanatory research design rooted in environmental economics and SSCM. The design is appropriate for analyzing preference structures, trade-offs, and behavioral responses to wetland conservation alternative financing options at a specific point in time. The study was done in SW on the eastern coast of Peninsular Malaysia (Figure 10.1). The study area was chosen for its ecological, economic, and conservation value.

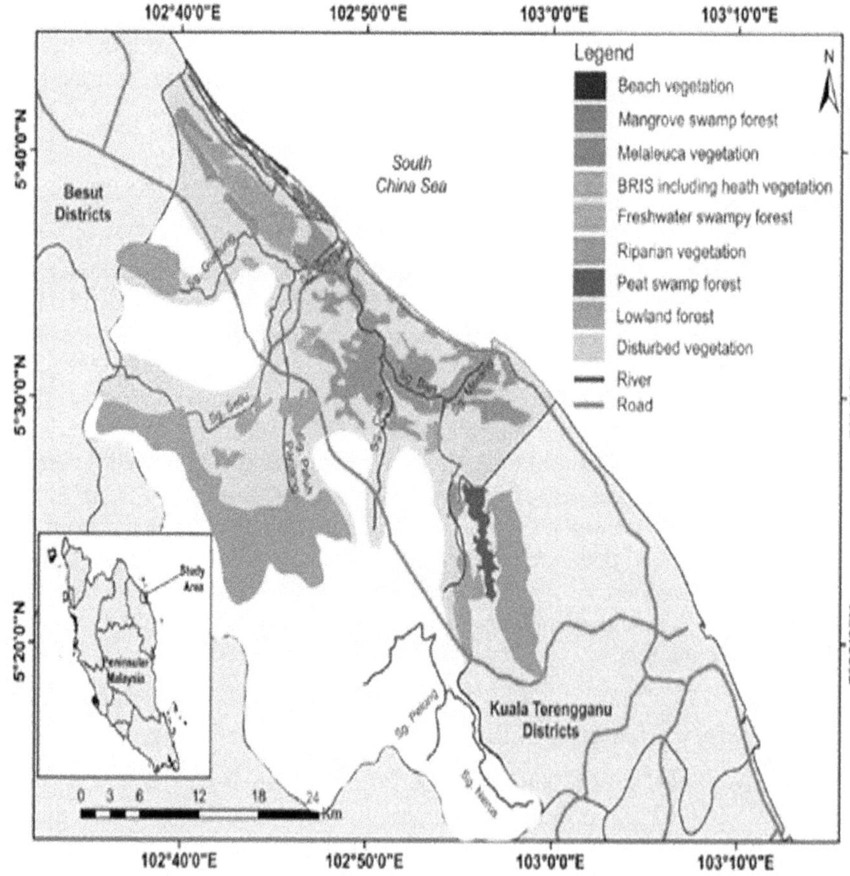

FIGURE 10.1 Location of SW.

(Image taken from Salim et al., 2020.)

TABLE 10.1
Attributes, Level, and Sources

Attribute	Levels	Source (Literature/Focus Groups)
Biodiversity	Status quo, medium, high	Barbier (2013); focus group validation
Flood risk reduction	Status quo, medium, low	Community input; focus group validation
Buffer zone	Status quo, medium, high	Lee and Du Preez (2016), Newton (2012); focus group validation
Recreational services	Status quo, medium, high	Turner et al. (2011); focus group validation
Payment mechanism	Donation, tax, subsidy reduction (dummy)	Hassan et al. (2018), Poudel et al. (2020), Siew et al. (2015); focus group validation

10.3.1 Choice Experiment Attributes and Levels

The attributes and levels were determined from pertinent literature and initial focus group discussions with local stakeholders. Participants in focus groups included five village leaders from the SW areas, members from the Municipal Council of Setiu, and a representative from WWF Setiu. Their feedback validated attributes and level selection to ensure that the CE represented scientific relevance and local socioeconomic realities. Table 10.1 shows the final set of attributes, levels, and sources. An example of choice sets is presented in Figure 10.2.

The CE technique was chosen for its ability to encapsulate trade-offs among various wetland characteristics, which is crucial for assessing ecosystem services as well as connecting the ESV with SSCM model. In contrast to contingent valuation, CE facilitates the differentiation of respondents' preferences.

10.3.2 Data Collection

Data was collected using a structured questionnaire administered to 2,000 participants via stratified sampling, stratified by payment mechanism. Respondents were randomly assigned to one of the three payment mechanism treatments to ensure comparability and internal validity across groups, even as this approach balances randomization with experimental control, which is critical for behavioral valuation studies. A pilot test was conducted to refine the questions before actual data collection. Consequently, a total of 1,137 valid data points were obtained as the actual sample, yielding a 56.9% response rate and minimal survey attrition after initial engagement, in a controlled field setting. The survey was conducted to improve reliability, facilitate understanding of choice sets, and reduce non-response bias, especially in rural areas. The present study involves adults aged 18 years and older who are Malaysian residents and are familiar with, or rely indirectly on, wetland-related goods or services. Participants should also be able to understand and complete the survey in person. Exclusion criteria include incomplete or inconsistent survey, aged under-18s, and people who do not understand choice tasks after explanation.

Attributes	Policy 1	Policy 2	Current condition (SQ)
Buffer zone	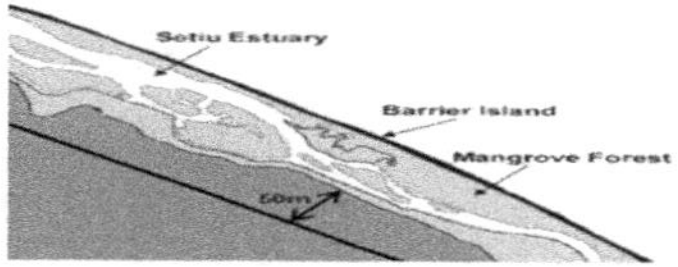 **Medium:** Protecting marsh habitat with 50 m buffer.	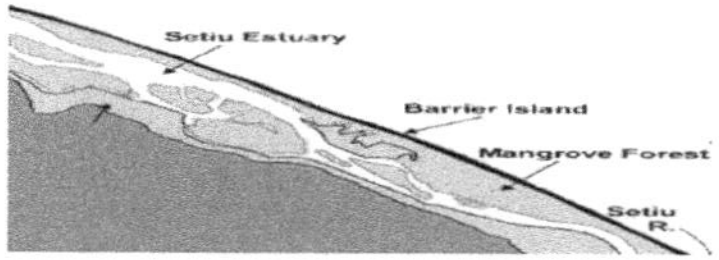 **Low:** The wetland habitat has no protection	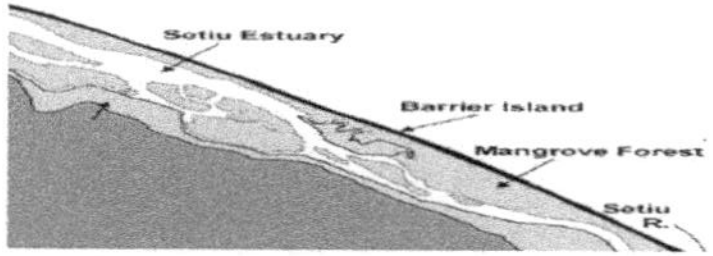 **Low:** The wetland habitat has no protection
Biodiversity enhancement	 **Low:** Few wetland species remain.	 **Medium:** A moderate population and species count.	 **Low:** Few wetland species remain.
Recreational quality	 **Low:** The recreation facilities are neglected, and there are few activities	 **Medium:** Possible for other facilities and activities.	 **Low:** The recreation facilities are neglected, and there are few activities
Flood risk control	 **Medium:** Wetlands may retain floodwaters to prevent fl ash floods.	 **Low:** Low risk and safe from flood.	 **High:** High risk, frequent flash/ heavy flood.
Payment (Donation)	**RM 5 /year**	**RM 210 /year**	**RM 0 /year**
I choose:	❑	❑	❑

FIGURE 10.2 Example of choice set the voluntary payment set presented to respondents.

10.3.3 Data Analysis and Model Estimation

The study applied a mixed logit model to analyze respondents' preferences for wetland conservation attributes using Biogeme software. The CE-derived utility function helps determine the marginal WTP for each attribute.

10.4 RESULTS

Although respondents were sampled from the public, the study is relevant to sectors whose supply chains rely directly or indirectly on wetland ecosystems, including tourism and hospitality (eco-tourism, coastal resorts); fisheries and aquaculture; agriculture and food supply chains; and local services, as well as informal wetland-based livelihoods. Therefore, the results are interpreted within a multisector SSCM perspective rather than through a single-industry lens.

10.4.1 Sociodemographic Results

Table 10.2 presents descriptive data on demographic characteristics across the three payment mechanism treatments. Notwithstanding the random allocation of participants across samples, the chi-square tests reveal significant differences by residential area and educational attainment. The educational level in the donation sample

TABLE 10.2
Distribution of Sociodemographics Across Different Payment Mechanisms

	Donation		Subsidies		Tax			
	Number	%	Number	%	Number	%	χ^2	p-value
Total respondent	411	36.1	403	35.4	323	28.4		
Residence area							20.68	0.00*
Urban	296	72	299	74.2	192	59.4		
Rural	115	28	104	25.8	131	40.6		
Gender							0.38	0.98
Male	202	49.1	200	49.6	161	49.8		
Female	209	50.9	203	50.4	162	50.2		
Education level							6.18	0.04**
Low level	296	72	258	64	215	66.6		
High-level	115	28	145	36	108	33.4		
Income							2.64	0.85
<RM12,000	180	43.8	174	43.2	156	48.3		
RM12,000–M24,000	132	32.1	130	32.3	96	29.7		
RM24,000–M36,000	52	12.7	52	12.9	34	10.5		
>RM36,000	47	11.4	47	11.7	37	11.5		

* Denote parameter is statistically significant at the 1% level.
** Denote parameter is statistically significant at the 5% level.

appears to be significantly lower than that of the other two groups. No significant difference was found in gender and economic distributions among the three payment mechanisms. The income level is particularly significant as theory dictates that income should influence WTP (Maleknia, 2024).

10.4.2 Wetland Preferences Aligning with SSCM

The results in Table 10.3 indicate that flood risk mitigation and biodiversity enhancement are the most significant factors influencing preferences. These findings indicate that biodiversity conservation is valued for local ecosystem and emphasize its significance within SSCM policies. Respondents regarded medium flood risk as equally significant as low flood risk, indicating local ecological awareness that seasonal flooding enhances soil fertility and fishery activities. Contrary to traditional beliefs that "less risk is invariably preferable," respondents attributed equal or greater value to moderate flood risk. In the context of SSCM, this suggests that conservation efforts should focus on controlling ecological variability rather than striving for its complete elimination to maintain ecosystem-dependent supply chains. Marriott International Serve 360 is committed to alleviating flood risks to hotel infrastructure, promoting biodiversity, and ensuring that operational resilience and brand integrity

TABLE 10.3
Preference Estimates (Mixed Logit Model)

Attribute	Coefficient	Std. Error	Interpretation
ASC (status quo)	0.485*	0.053	Respondents prefer current condition of the wetlands
Biodiversity – high	0.417*	0.071	High biodiversity strongly valued
Biodiversity – medium	0.093	0.068	Not significantly different from SQ
Buffer zone – high	−0.275*	0.063	High buffer zones perceived negatively
Buffer zone – medium	0.045	0.056	Not significantly different from SQ
Flood risk – low	1.010*	0.076	Strongly valued
Flood risk – medium	1.040*	0.064	Surprisingly valued as highly as low flood risk
Recreation – high	−0.338*	0.066	Negative utility for over-commercialized recreation
Recreation – medium	0.165**	0.067	Moderate improvements supported
Price – donation	0.0020*	0.0004	Positive acceptance
Price – tax	0.0015*	0.0005	Positive acceptance
Price – subsidy	−0.0110*	0.0004	Strong negative disutility
Final log-likelihood	−11,333.97		
Adjusted R^2	0.243		
Observations	1137		

* Denote parameter is statistically significant at the 1% level.
** Denote parameter is statistically significant at the 5% level.

indirectly showed support for wetland conservation (Marriott International, 2021; Wang et al., 2022).

Respondents opposed extensive commercialization of recreational activities, but endorsed moderate, environmentally sustainable facilities for recreation attributes. The negative response to wide buffer zones indicates local concerns around land-use limitations. Medium buffer zones received neutral to moderate supports proving that community in the conserved parks favors strategies that less interfere with their livelihoods. Report in Marriott International (2021) stated that community needs must be fulfilled before expanding the business. As highlighted by Mugoni et al. (2024), the SSCM practices must encounter both social needs and industry performance. Reduction in subsidies elicited a significant negative reaction among respondents and indicates that coercive measures are ineffective in emerging nations. The voluntary donation initiatives and green bonds for ecosystem restoration demonstrate how Marriott's implementation integrates voluntary and hybrid financing mechanisms into SSCM strategies, aligning consumer willingness with corporate sustainability objectives. Collectively, these findings indicate that SSCM must be flexible, integrating conservation finance, sustainable development, and land-use strategies that correspond with ecological and economic objectives.

10.4.3 Mean WTP Across Different Payment Mechanisms

Table 10.4 presents the mean WTP for wetland conservation across donations, taxes, and subsidy reductions. Donation programs exhibit the highest WTP, succeeded by taxation and potential negative impacts from subsidy reduction. WTP for low flood risk is RM 132.24/year through donation, RM 104.93 through taxes, and RM 87.76 via subsidy reductions. In terms of payment through donation (RM 67.65), the valuation of biodiversity protection surpassed tax (RM 44.70) or subsidy reduction (RM 20.68). Conversely, extensive recreational development and buffer zones were associated with negative WTP across all payments. Figure 10.3 illustrates the differences in the financing mechanism. The results support hypotheses H2 and H3, indicating that coercive strategies are less valuable than voluntary or government-supported alternatives.

TABLE 10.4
The Mean WTP for Split Payment Vehicle (RM/Year)

Attribute	Donation	Tax	Subsidy
Biodiversity – high	67.65	44.70	20.68
Flood risk – low	132.24	104.93	87.76
Flood risk – medium	115.85	130.12	91.98
Recreation – medium	33.99	19.83	0.50
Recreation – high	−17.05	−48.69	−71.52
Buffer zone – high	−3.18	−35.05	−50.74

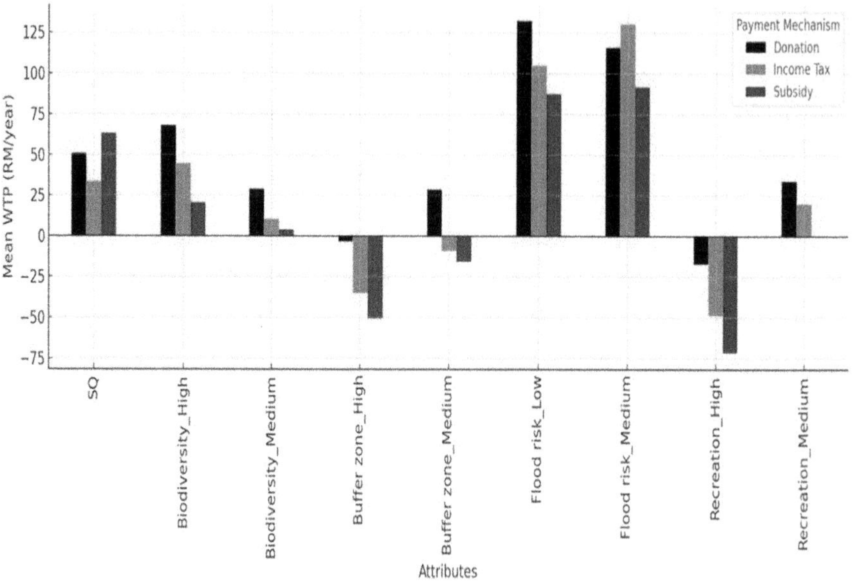

FIGURE 10.3 Mean WTP based on different payment mechanisms.

10.5 DISCUSSION OF FINDINGS

The findings highlight the importance of wetland valuation in informing SSCM, especially regarding flood risk mitigation and biodiversity enhancement, which are highly valued. This discovery corroborates previous ecosystem valuation research, which consistently demonstrates that regulatory and supporting services yield greater welfare benefits than cultural or recreational services and boost public well-being (Barbier, 2013; Dingha et al., 2025; Ibrahim et al., 2025; Turner et al., 2011). This improves the SSCM literature, which has focused on emissions reduction and logistical efficiency rather than on ecological dependency (Gupta & Palsule-Desai, 2011; Mugoni et al., 2024; Olaleye et al., 2025; Shekarian et al., 2022). A major finding is that medium flood risk is as favored as low flood risk, undermining the supply chain and risk management facts that risk minimization is always best. This study confirms ecological data indicating that moderate flooding improves soil fertility, fish spawning, and nutrient cycling, which are important for wetland-based livelihoods (Salimi et al., 2021). Participants favored low-impact recreational facilities, supporting Arnegger et al.'s (2024) ecotourism and mass tourist avoidance. Excessive infrastructure development may benefit in the short term, but harm the environment and supply chain. Buffer zone preference corroborates previous research indicating resistance to stringent conservation zoning when livelihoods depend on ecosystem access (Lee & Du Preez, 2016; Oliveira & Meyfroidt, 2022). The donation-based and tax-based financing systems substantially increase respondents WTP, whereas reductions in subsidies produce considerable negative utility. This analysis enhances PES literature by empirically evaluating three payment vehicles within an SSCM setting, which has predominantly analyzed payment mechanisms in isolation (Chen et al., 2014; Cranford & Mourato, 2014).

10.6 CONCLUSION AND IMPLICATIONS

10.6.1 Summary of Findings

The benefits of wetland are not always included in supply chain decision-making. This omission leaves industries like tourism and fishing vulnerable to environmental risks. The main findings confirm that conservation attributes significantly influence preferences. Flood risk reduction and biodiversity improvement were highly valued. Medium flood risk was unexpectedly preferred over low risk, indicating that earning a livelihood is more important in that area. Large-scale recreational commercialization was rejected, but moderate improvements received support. The creation of buffer zones was less popular, indicating concerns about land-use restrictions. Finally, donation and tax-based funding sources were favored over reducing household subsidies.

10.6.2 Practical Implications and Theoretical Contributions

This study adapts SSCM theory to include wetland ecosystem services as supply chain resilience inputs. Favoring medium flood risk over environmental risk may boost productivity and performance in managed ecological variability. The paper also contrasts the mechanisms of donation, tax reduction, and subsidy reductions in an integrated valuation system whereby, the economic capacity, institutional legitimacy, and perceived coerciveness play a role in influencing WTP enriching behavioral and institutional means of environmental valuation. Wetland conservation forms part of the risk and sustainability management of tourism, fisheries, agriculture, and coastal infrastructure. Corporations should adopt moderate- and low-impact approaches to ensure stability of supply chains because they are compliant with environmental and social standards. Overall, as the study points out, conservation financing, ecological health, and supply chain governance are the prerequisites of sustainable development and resilient supply chains.

10.6.3 Limitations of the Study and Suggestions for Future Research

This study sheds light on SSCM wetland conservation valuation with some limitations. A stated-preference CE may not reflect actual behavior, causing hypothetical bias. For further realism, future study might use/reveal preferences to compare with this CE output. The current study focuses on SW in Malaysia that limits its applicability. Multiple wetland comparisons may increase external validity. Extended conservation criteria and funding choices might better reflect altering funding systems than the study's limited. Finally, this SSCM models focus on public preferences rather than company and supply chain viewpoints, so stakeholder-driven decision-making frameworks could improve conservation policy implementation in the future.

ACKNOWLEDGMENTS

The authors sincerely thank B.J. Thorsen and S.B. Olsen for their invaluable support and guidance throughout this project.

AUTHORS' DECLARATION

The information and data provided in the manuscript do not present any conflicts of interest and also strictly adhere to and comply with all ethical publication procedures.

REFERENCES

Arnegger, J., Herz, M., & Campbell, M. (2024). Mass ecotourism, media, and wildlife experience. *Journal of Outdoor Recreation and Tourism*, 45, 100732.

Barbier, E. B. (2013). Valuing ecosystem services for coastal wetland protection and restoration: Progress and challenges. *Resources*, 2(3), 213–230.

Chen, X., Viña, A., Shortridge, A., An, L., & Liu, J. (2014). Assessing the effectiveness of payments for ecosystem services: An agent-based modeling approach. *Ecology and Society*, 19(1). https://doi.org/10.5751/es-05578-190107

Cranford, M., & Mourato, S. (2014). Credit-based payments for ecosystem services: Evidence from a choice experiment in Ecuador. *World Development*, 64, 503–520.

Dharmayanti, N., Ismail, T., Hanifah, I. A., & Taqi, M. (2023). Exploring sustainability management control system and eco-innovation matter sustainable financial performance: The role of supply chain management and digital adaptability in Indonesian context. *Journal of Open Innovation: Technology, Market, and Complexity*, 9(3), 100119.

Dingha, C. B., Biber-Freudenberger, L., Mbanga, L. A., & Kometa, S. S. (2025). Community-based valuation of wetland ecosystem services: Insights from Bamenda, Cameroon. *Wetlands Ecology and Management*, 33(1), 14. https://doi.org/10.1007/s11273-024-10029-w

Gupta, S., & Palsule-Desai, O. D. (2011). Sustainable supply chain management: Review and research opportunities. *IIMB Management Review*, 23(4), 234–245.

Hassan, S., Olsen, S. B., & Thorsen, B. J. (2018). Appropriate payment vehicles in stated preference studies in developing economies. *Environmental and Resource Economics*, 71(4), 1053–1075.

Hein, L., Miller, D. C., & de Groot, R. (2013). Payments for ecosystem services and the financing of global biodiversity conservation. *Current Opinion in Environmental Sustainability*, 5(1), 87–93.

Hejnowicz, A. P., Raffaelli, D. G., Rudd, M. A., & White, P. C. L. (2014). Evaluating the outcomes of payments for ecosystem services programmes using a capital asset framework. *Ecosystem Services*, 9, 83–97.

Ibrahim, M. S. N., Johari, S., Assim, M. I. S. A., Mohammad, S. K. W., Afandi, S. H. M., & Hassan, S. (2025). An economic assessment on public well-being for biodiversity conservation in Gunung Mulu National Park, Sarawak, Malaysia. *Environment, Development and Sustainability*, 27(6), 14497–14521.

Le, T. A. T., Vodden, K., Wu, J., Bullock, R., & Sabau, G. (2024). Payments for ecosystem services programs: A global review of contributions towards sustainability. *Heliyon*, 10(1), e22361.

Lee, D. E., & Du Preez, M. (2016). Determining visitor preferences for rhinoceros conservation management at private, ecotourism game reserves in the Eastern Cape Province, South Africa: A choice modeling experiment. *Ecological Economics*, 130, 106–116.

Lola, M. S., Ramlee, M. N. A., Isa, H., Abdullah, M. I., Hussin, M. F., Zainuddin, N. H., & Rahman, M. N. (2016). Forecasting towards planning and sustainable development based on a system dynamic approach: A case study of the Setiu District, State of Terengganu, Malaysia. *Open Journal of Statistics*, 6(5), 931–950.

Maleknia, R. (2024). Psychological determinants of citizens' willingness to pay for ecosystem services in urban forests. *Global Ecology and Conservation*, 54, e03052.

Marriott International. (2021). *Serve 360: Doing Good in Every Direction – Sustainability and Social Impact Report.* Marriott International. https://serve360.marriott.com/ Accessed on 25th August 2025.

Mugoni, E., Kanyepe, J., & Tukuta, M. (2024). Sustainable Supply Chain Management Practices (SSCMPS) and environmental performance: A systematic review. *Sustainable Technology and Entrepreneurship*, 3(1), 100050.

Newton, G. (2012). Buffer zones for aquatic biodiversity conservation. *Journal of the Australian Network for Plant Conservation*, 2(21), 18–22.

Olaleye, B. R., & Mosleh, S. F. (2025). Greening sustainable supply chain performance: The moderating and mediating influence of green value co-creation and green innovation. *Administrative Sciences*, 15(5), 183.

Olaleye, B. R., Sekhampu, T. J., Babatunde, B. O., Hassan, S., Efuntade, A. O., & Bamidele, J. A. (2025). Going green: Effect of dynamic capability, innovation, and supply chain agility in fostering supply chain management in SDGs context. *Journal of Lifestyle and SDGs Review*, *5*(3), e05056.

Oliveira, E., & Meyfroidt, P. (2022). Strategic spatial planning in emerging land-use frontiers: evidence from Mozambique. *Ecology and Society*, 27(2).

Poudel, R., Collins, A., Gazal, K., & Wang, J. (2020). Benefit transfer estimation of willingness-to-pay for U.S. wetlands conservation. *Forest Policy and Economics*, 115, 102157.

Salimi, S., Almuktar, S.A.A.A.N., and Scholz, M. (2021). Impact of climate change on wetland ecosystems: a critical review of experimental wetlands. *Journal of Environmental Management*, 286, 112160.

Salim, J. M., Lee, G. E., Salam, M. R., Shahimi, S., Pesiu, E., Jani, J. M., Horsali, N. A. I., Shahrudin, R., Nor, S. M. M., Chong, J. L., Mohamad, F., Raffi, A., & Nikong, D. (2020). A checklist of vascular plants and uses of some species for livelihood-making in Setiu Wetlands, Terengganu, Malaysia. *PhytoKeys*, 160, 7–43.

Salmi, A., Quarshie, A. M., Scott-Kennel, J., & Kähkönen, A. K. (2023). Biodiversity management: A supply chain practice view. *Journal of Purchasing and Supply Management*, 29(4), 100865.

Shekarian, E., Ijadi, B., Zare, A., & Majava, J. (2022). Sustainable supply chain management: A comprehensive systematic review of industrial practices. *Sustainability*, 14(13), 7892.

Siew, M., Yacob, M., Radam, A., Adamu, A., & Alias, E. F., (2015). Estimating willingness to pay for wetland conservation: A contingent valuation study of Paya Indah Wetland, Selangor Malaysia. *Procedia Environmental Sciences*, 30, 268–272.

Tchonkouang, R. D., Onyeaka, H., & Nkoutchou, H. (2024). Assessing the vulnerability of food supply chains to climate change-induced disruptions. *Science of the Total Environment*, 920, 171047.

Turner, R. K., Georgiou, S. G., & Fisher, B. (2011). *Valuing Ecosystem Services: The Case of Multi-functional Wetlands.* Earthscan.

Tzanidis, T., Magni, D., Scuotto, V., & Maalaoui, A. (2024). B2B green marketing strategies for European firms: Implications for people, planet and profit. *Industrial Marketing Management*, 117, 481–492.

VanderWilde, C. P., Newell, J. P., Gounaridis, D., & Goldstein, B. P. (2023). Deforestation, certification, and transnational palm oil supply chains: Linking Guatemala to global consumer markets. *Journal of Environmental Management*, 344, 11805.

Wang, X., Prahalad, V., & Kirkpatrick, J. B. (2022). Public perceptions of wetlands and preferences for on-site visitor facilities and communication media: A case study from an Australian Ramsar wetland. *Marine and Freshwater Research*, 73(10), 1149–1165.

11 An Integrated Approach to Improved Value-Chain System for Jute in Bangladesh

A Ground-Level Investigation

Sonia Shahid, Md. Sohel Rana, and Md. Khaled Saifullah

11.1 INTRODUCTION

Bangladesh has been relying on jute as an important agricultural and industrial product because it is known as the Golden Fiber. The jute value chain in Bangladesh experiences profound problems, which limit its relevance, success, and justice. Conventional agricultural activities are labour intensive, which contributes to the high cost of production and labour scarcity, and the lack of attraction to young people. This study applies two theories that complement one another, which are the value chain theory (Kaplinsky & Morris, 2000) and the sustainable livelihoods framework (DFID, 1999). The value chain theory can be used to examine how governance, coordination, and value distributions occur in the upstream (farming), midstream (processing), and downstream (design, branding, export) sectors of jute by showing that systemic inefficiencies and asymmetries of power existed. The European Union has increased the need for biodegradable alternatives, which projects a global jute bag market of 500 billion units per year (Asaduzzaman, 2022). However, Bangladesh is not able to capitalize because of low productivity and value addition. Conversely, India, the largest producer of jute in the world, has been spending heavily on the mechanization of decortication, design innovations and policies that make jute mandatory in India.

The recent empirical tendencies highlight the necessity of revitalization. Between 2019 and 2020, jute and jute diversified products (JDPs) have added more than 20% to the increase in export earnings (BIDA, 2023), but the trend has slowed down. The price of domestic labor doubled or even tripled between BDT 300 and BDT 800 per day in a year (2021–2022) (Joshi, Panagiotou & Rahman, 2024), making it no longer

DOI: 10.1201/9781003770008-16

an option to use manual labor to harvest crops among smallholders. The jute industry of Bangladesh faces many challenges, such as low productivity, old infrastructure, inability to innovate, and poor coordination between the jute farmers, processors, designers, exporters, and policymakers (Gebresenbet & Oodally, 2005). This is further aggravated by a lack of mechanization, access to credit and integration of agricultural practices with market design orientation. There are also some studies on the jute value chain in Bangladesh, but these studies remain stagnant without unity. Even though Barua and Rahman (2019), Barua, Rahman, and Barua (2021), and Rifath (2018) mention aspects of the dynamics of the supply chain and sustainability, they are denounced, respectively, to discuss technological adoption, the engagement of youth, innovation in design, and financial inclusion in one coherent research.

The key questions that will be answered by this research will include how a workplace can be evaluated prior to and following an improvement intervention. What might be considered the major obstacles to a more efficient and expansive jute value chain in Bangladesh? And what role can be played by policy refreshments and institutional cooperation in the re-emergence of the jute industry and inspiring youth agitation?

The chapter presents a detailed strategy toward the empowerment of the Bangladesh jute value chain. It examines existing literature and describes the research methodology, results on four topics, including limits to labor and mechanization, inadequate youth participation and training, finance and policy disconnect, and design-driven innovation opportunities. The discussion provides an integrated model of interaction between enterprise innovation, equitable financing, and modernization of farming, and, finally, conclusion and recommendations.

11.2 LITERATURE REVIEW

In order to combine theoretical understandings and empirical realities in the jute industry of Bangladesh, this chapter incorporates the value chain theory (Kaplinsky & Morris, 2000) and sustainable livelihoods framework (DFID, 1999) into one-concept model. The use of value chain theory assists in unbundling governance, power, and distribution of values among upstream (farmers), midstream (processors), and downstream (designers, exporters). In 2019–2020, the export of jute products jumped to over 20% and the global demand for jute shopping bags currently stands at 500 billion pieces (Asaduzzaman, 2022). There is a strong need for concerted efforts in supporting farm mechanization for jute and market development (both local and export) through green financing and small business development (Hossen, 2019). Despite these fluctuations, the BBS (2019) recorded a healthy growth discernible in the growth of jute and JDPs exports by 20% between 2019 and 2020.

Different sections of the jute value chain have been investigated by various studies. Barua and Rahman (2019) examined sustainable values in agricultural chains, emphasizing the importance of the coordination of stakeholders. Barua, Rahman, and Barua (2021) investigated livelihoods based on livestock in coastal Bangladesh and emphasized the role of the nexus with institutions. In a study by Rifath (2018), there was a mention of a lack of efficiency in logistics and engagement with the farmers in the upstream jute supply chain due to a conceptual model made for the upstream

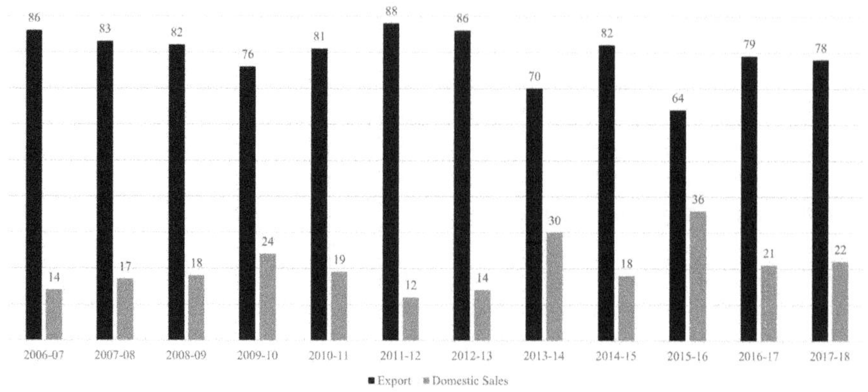

FIGURE 11.1 Quantity of exports and domestic sales of jute and JDP (% of total sales). (BBS, 2019.)

jute supply chain. Simultaneously, a dichotomous value chain analysis was conducted on jute fiber by Moniruzzaman et al. (2024) and identified bottlenecks in the processing and export competitiveness. Conventional farming is based on manual input using hand tools or animal drafts, which implies high levels of "drudgery." Small-scale farming tends to possess these characteristics: "poor, insecure returns; physically damaging/demanding conditions; and/or illegal status" (Petesch & Caillava, 2012) (Figure 11.1).

Mechanization service providers (MSPs) provide equipment and machinery rentals to about 100 farmers each and recover the cost of both the machinery/equipment and its maintenance. Agricultural MSPs frequently work on a seasonal basis and concentrate on harvesting and land preparation (plowing). As agricultural field operations move from manual to fully mechanized processes, the employment of the laborers will also shift from seasonal or temporary to full-time employment (Kirsch, 2018). Growing labor-intensive industry, low level of mechanization, disaggregated components of the supply chain, and declining interest among younger generations are some of the issues that affect the sector.

The jute agriculture is highly labor intensive, especially at harvesting and resting. This creates a lot of hard work and a lack of youth participation (Petesch & Caillava, 2012; Hossen, 2019). The research indicates that there are policy implementation gaps, uncourageous support services, along with lack of credit options among smallholder farmers (ADB, 2016; Kabir & Islam, 2021). The lack of a unified regulatory framework worsens the development of the sector as well. The global market demand is growing toward the usage of eco-friendly materials, and in this situation, Bangladeshi jute is competing tightly with synthetic materials, and it has no branding or certification as an eco-friendly resource (Azad et al., 2020). Jute helps millions of people in rural areas. Nevertheless, falling profitability and environmental degradation through the traditional retting techniques put its sustainability at stake (Ali et al., 2015). Such studies rely on value chain theory (Kaplinsky & Morris, 2000), which emphasizes coordination, value-addition, and value enhancement at various stages of the production process.

This review has three research gaps: (1) within value-chain models: absence of integrated models that connect farming, processing, design, and markets; (2) no evidence of how technology and skills development can mobilize the youth amid labor shortages; (3) policy-practice gap: little evidence on inter-agency collaboration or implementation. The external factors comprise the irregularities in the market, governance problems in BJMC, governmental deprioritization, and anti-dumping duties introduced in India in 2017, which significantly decreased the jute exports (New Age, 2020, June 30).

11.3 METHODOLOGY

The study implements the qualitative research method in order to find the challenges and opportunities within the value chain of the jute industry in Bangladesh. The key aim is to develop contextual, detailed insights into the critical opinion seekers on the nature of technological adoption, activities with young people, policy constraints, and innovation in JDPs. This research project does a qualitative case study of the systematic issues and opportunities of the jute value chain system in Bangladesh by using an exploratory, interpretive, qualitative research design.

11.3.1 Data Collection Methods

Primary data were collected through semi-structured in-depth interviews and field observations during the third quarter of 2023. A total of 15 key informants were interviewed, representing diverse segments of the jute value chain, including five jute farmers, two agriculture officers and extension agents, seven jute research scientists and promotion officers, and one bank manager involved in agricultural financing. Purposive sampling was used to select participants so that all these areas were represented in the sample, such as individuals with direct involvement and knowledge of jute cultivation, processing, policy, and financing. The semi-structured interview guide was developed with 11 open-ended questions in this study (Table 11.1).

The participants were selected based on their current engagement and on how they have driven the social, economic, technological, and institutional elements in the jute and agriculture sectors. The respondents were from different jute-related government institutions, the private sector, farmers, traders, NGOs, CSOs, and research institutes dedicated to the jute sector.

11.3.2 Data Analysis

The audio-recording of all the interviews (with consent) was followed by transcription of the interviews verbatim and the thematic analysis of the results with the help of Braun and Clarke (2006). Six stages underwent the analytical process. Acclimatization with the data was done through repeated reading of transcripts. Preliminary coding of major statements associated with the value-chains, challenges and opportunities, coming up with initial themes (e.g., labor drudgery, financing

TABLE 11.1
Sample Category

Sample	Category
Producer and traders	Jute farmer Mechanization service providers to the jute farmers Jute and jute stick trader Jute seed seller
Jute manufacturer	Jute mills association Charcoal factory entrepreneur Partex particle board manufacturer Akij Particle Board Ltd
Machinery distributor	The Metal Pvt. Ltd. RK Metal Nasim Engineering Works Tahura Engineering Workshop
Research & development	CSISA-MEA (2022) Bangladesh Agricultural Research Institute Bangladesh Jute Research Institute
Policymakers and implementors	Department of Agricultural Extension (Mechanization Project) Bangladesh Krishi Bank Probashi Kallyan Bank Upazila Krishi Officer/Agriculture Officer Department of Agricultural Marketing

gaps, and defunct design-market), revision and correction of themes to have internal consistency and unicity, and final themes are named. Finally, themes are described by quoting the representative verbatim.

11.3.3 Demographic Discussion

Table 11.2 shows that out of 15 respondents, 86.7% are male, and the remaining are female. In terms of education level, the majority of the respondents (66.7%) have a post-graduation level of education. The table also shows that the majority of the respondents are jute farmers, followed by jute promotion officers, jute research officers, agriculture officers, and bank managers. Moreover, the average age of the respondents was 30 years, and the average year of experience was 12.

11.3.4 Maintaining Goodness of Fit and Trustworthiness

Credibility: Superior validity was achieved through prolonged involvement with the participants, checks by the members (presenting the initial results of the research to two respondents), and triangulation of the data (interviews and observations). *Transferability*: Profiling of participants and research setting and thick description of context provide transferability to other rural agricultural systems. *Dependability*:

TABLE 11.2
Respondents' Characteristics

Measure	Items	Frequency	Percentage
Gender	Male	13	86.7
	Female	2	13.3
Education	Primary school	4	26.7
	High school	1	6.6
	Post-graduation	10	66.7
Occupation	Agriculture officer	2	13.3%
	Bank manager	1	6.7%
	Jute farmer	5	33.3%
	Jute promotion officer	4	26.7%
	Jute research officer	3	20.0%
	Average	**Maximum**	**Minimum**
Age	39	49	27
Years of experience	12	20	3

Source: Author.

A detailed audit trail was provided that tracked all steps in data collection, coding decisions, and theme development. *Reflexivity*: Reflexivity was applied in the research process, trying to reduce the bias of the research. A research journal was maintained with personal thoughts as well as assumptions (Figure 11.2).

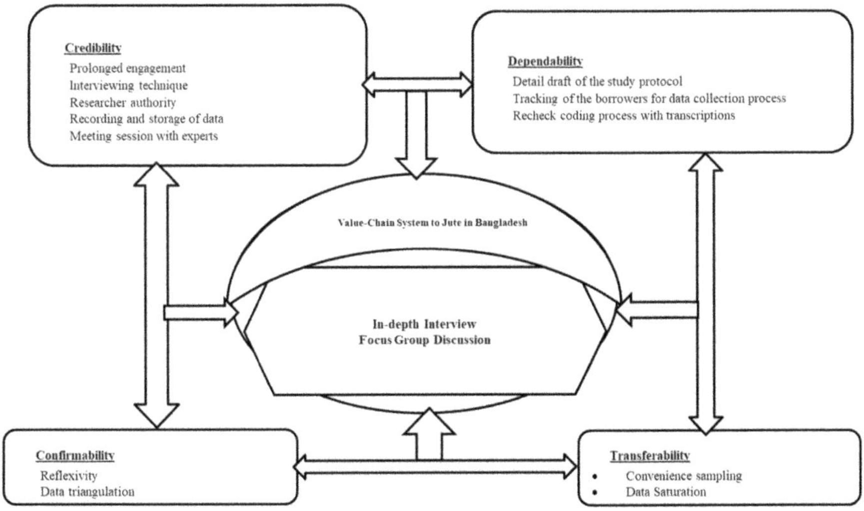

FIGURE 11.2 Goodness of fit and trustworthiness.

(Guba & Lincoln, 1985.)

11.3.5 Ethical Considerations and Limitations

The consent of all participants was taken before the interviews. It guaranteed participants' anonymity and confidentiality and reported information was stripped of identifiers. No monetary rewards were offered, and subjects were told that they could withdraw at any moment.

11.4 FINDINGS

The results show severe issues and emerging opportunities in the jute value chain of Bangladesh. The findings are organized around five general spheres. That is, labor and mechanization, youth participation and skills building, access to finance and policy facilitation, design innovation and market connections, and institutional coordination. Indicatively, 86.7% of interviewees cited the labor shortage as a major limitation to the process and production of jute. Also, Table 11.3 presents the levels of stakeholder awareness of the existing jute-related policies (e.g., 32% were only aware of the National Jute Policy 2023) and Table 11.4 shows patterns and barriers to access

TABLE 11.3
Respondents' Profile

Indicator	Findings	Source
Respondents	15 total: 5 farmers, 2 agriculture officers, 4 jute promotion officers, 3 research officers, and 1 bank manager	Field
Male respondents	13 (86.7%)	Field
Post-graduate education	10 (66.7%) among non-farmers	Field
Farmers with mechanized access	2 out of 5	Field
Farmers with formal loans	1 out of 5	Interviews
Awareness of export assistance schemes	7 out of 15 (47%)	Interviews

Source: Field Study.

TABLE 11.4
Policy Support Awareness of the Respondents

Policy Support Awareness	Aware	Not Aware
Agriculture officers (n = 2)	1 (50%)	1 (50%)
Jute officers (n = 7)	5 (71%)	2 (29%)
Farmers (n = 5)	1 (20%)	4 (80%)

Source: Authors.

institutional finance among various actors in the value-chain, showing that fewer than 20% of smallholders have received institutional credit over the last three years.

11.4.1 Labor and Mechanization Challenges

The jute industry is still very reliant on manual labor, especially when it comes to harvesting and retting, which increases the level of drudgery and inefficiency; 86.7% of the respondents (n = 13) cited labor scarcity as a key limitation. According to farmers, manual harvesting requires 3–4 days per acre, whereas mechanized harvesting requires 4–6 hours.

11.4.2 Young Participation and Capacity Building

The challenge of youth disengagement is increasing, and the reason is that jute farming is regarded as an old and ineffective line of work. All the respondents (n = 15) said that the youths are shunning jute farming. It was proposed by the respondents that vocational training in jute design, processing, and marketing would help to bring the youth to value-added jobs instead of fieldwork.

11.4.3 Financial and Policy Support

Policy fragmentation was also identified by the respondents: a variety of ministries (Agriculture, Textiles, and Commerce) are engaged; yet, there is a lack of coordination.

11.4.4 Market Connection and Design Innovation

There is a high disconnect between the jute producers and the international buyers. Big-volume clients (e.g., IKEA, Walmart) require eco-certified, design-oriented products, which Bangladeshi JDPs do not always have branded and consistent with their required demand. Only two of four Jute Promotion Officers were in direct contact with international buyers.

11.4.5 Institutional Coordination and Forward/Backward Linkages

Lack of scalability is due to weak coordination among institutions. There is no unified platform where farmers, researchers, designers, and exporters can interact. The respondents demanded a National Jute Development Authority to facilitate policies and investments. On field observations, jute sticks (by-products) are frequently burnt or discarded, even though they may be used as briquettes, handicrafts, or bioenergy. The model (Figure 11.3) shows that the scale of production should be accompanied by parallel investment in infrastructure and services.

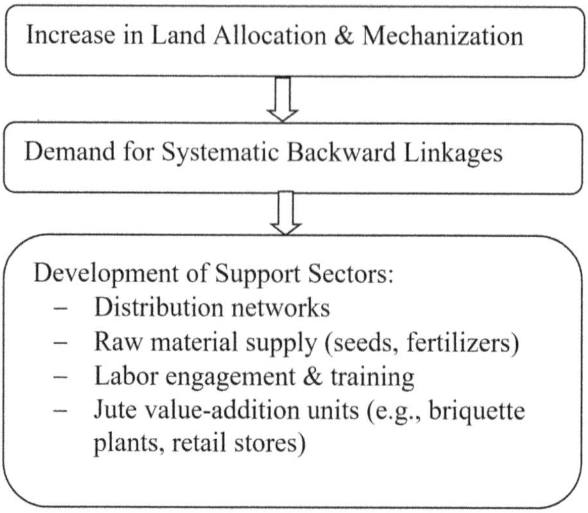

FIGURE 11.3 Systematic backward linkage development for jute value addition.

(Authors.)

11.5 DISCUSSION

Findings of this study show that the jute value chain in Bangladesh has systemic problems in the areas of labor, technology, finance, design, and institutional coordination. None of these issues exist in a vacuum, but in fact contribute to one another that lead to a chain where productivity is being hampered, the youth are being discouraged, and export competitiveness is hampered. In this section, the most important findings are discussed analytically with support of empirical data and locating them in a broader context of sustainable agriculture and changes in the countryside. The chapter combines the value chain theory and the sustainable livelihoods approach. Important empirical results have now been made comparatively with the existing literature, e.g., our results on limited mechanization in jute processing are compared with the findings of Hossen (2019) on agricultural mechanization in Bangladesh and our results on youth disengagement in jute processing are discussed in relation to the work of Petesch and Caillava (2012) on youth aspirations in the rural areas. In appropriate cases, differences are pointed out (e.g., our study, unlike Barua and Rahman (2019). The improvements make sure that the conversation is no longer descriptive, but analytical and academic in nature.

11.5.1 Work Intensity and Requirement of Mechanization

In the field observations and interviews, it can be noted that manual harvesting consumes 3–4 days on each acre of land. Whereas the mechanized process consumes less than 6 hours, which depicts a gap of more than 90% reduction. However, only 40% of the farmers interviewed obtained access to machines such as power tillers or decorticators. This deficit points to a larger structural problem: smallholders cannot get access to mechanization due to prohibitive costs of entry, poor access to credit, and the fact that machines are not available or are lacking suitable ones.

11.5.2 The Disengagement of Youth Is Accomplished Through Perceptions and Possibilities

All 15 stakeholders observed a decline in the youth involved in jute cultivation. However, whereas 66.7% of non-farmer respondents have post-graduate degrees (Table 11.2), only 6.6% of farmers have completed high school or higher. This is evidence of skills and perception gaps. The figures show that youth do not oppose agriculture, but do not prefer low-valued and hard labor of a farming routine that is not dignified and innovative.

This revelation is consistent with that of Petesch and Caillava (2012), who describe drudgery as one of the primary reasons behind youth disengagement in rural areas. The 86.7% male ratio among the respondents (Table 11.2) also indicates that jute farming is becoming a male-dominated and physically demanding job that is not appealing to young and better-educated people. Lack of mechanization is not just a technical gap but also a socioeconomic obstacle to intergenerational continuity in jute farming.

11.5.3 Financial Exclusion and Policy Fragmentation

The availability of finance proved to be a significant drawback. Fewer than five farmers (20%) had been granted a formal loan. The remaining are dependent on the informal lenders who charge an interest of 20–30% per year, contrary to the official rate of agricultural lending of 4–9% (Table 11.5).

The Bank Manager (BKBMGR) explained this by the fact that there were no collateral and land titles, which is a typical problem with smallholder farmers. In addition, the lack of policy coordination and the absence of implementation gaps are caused by policy fragmentation between ministries (agriculture, textiles, and commerce). The low-investment, low-productivity trap is a perpetuation of financial exclusion.

TABLE 11.5
Access to Finance

Received a Formal Loan		Relied on Informal Sources	
Frequency	**Percentage**	**Frequency**	**Percentage**
1	20	4	80
4	40	2	20

Source: Primary interviews, Q3 2023.

11.5.4 Design Innovation and Market Linkages: Bridging the Gap

The most startling fact is that producers and international buyers are not connected. The major retailers such as IKEA and Walmart require eco-certified and design-driven products. Nevertheless, Bangladeshi Jute Development Projects tend not to be branded, consistent, and market-smart. There were only two Jute Promotion Officers who were in direct contact with international buyers. None of them complained about frequent market feedback. The net effect of this is that there will be poor product development and a lack of export opportunities. The current value chain is linear and fragmented. It goes from farm to raw fiber export. To capture more value, it needs to become circular and integrated, where design leads production instead of the other way around. Moreover, privatization of jute mills and spinning is the right way forward to bring in efficiency gains. In addition, the mills can link to the garment manufacturers, where the efficiency gains can be shared between the two sectors.

11.5.5 The Need for Labor Efficiency

With the current trend in jute production and projected demand, the total expected production in 2022 will need to stand at 1600 tons of jute fiber (CEIC, 2023), which equates to a total land area of 700,000 hectares engaged in production. This requires massive engagement of farmers and farmland through youth development programs and training dissemination through public and private training institutes. Labor efficiency is influenced by safe job assurance and high workmanship.

11.5.6 Training Programs on the Jute Cultivation Process

The training curricula require much greater collaboration and technical assistance with the NGOs and researchers to develop a simplified and practical approach to incorporating the use of new technologies that are being developed. Additionally, effective intercropping and an enhanced retting procedure boost jute farming's profitability. Therefore, implementing these technologies will meet the need for jute fiber, boost jute farmers' incomes, and protect the environment (Kumar et al., 2014). Farm training plays a big part in encouraging and ensuring the adoption of better jute production techniques. Training has a significant impact on the knowledge and behavior of beneficiary farmers compared to non-beneficiary farmers (Sadat, Ghosh & Chakraborty, 2017).

11.5.7 Toward an Integrated Value-Chain Framework

The data support the development of an Integrated Jute Value-Chain Model (Figure 11.4), which connects: (1) Upstream: Mechanization, training, and youth engagement. (2) Midstream: Processing efficiency, retting innovation, and quality control. (3) Downstream: Design-led product development, branding, and global market access. (4). Enablers: Policy coordination, financial inclusion, and research-extension linkages.

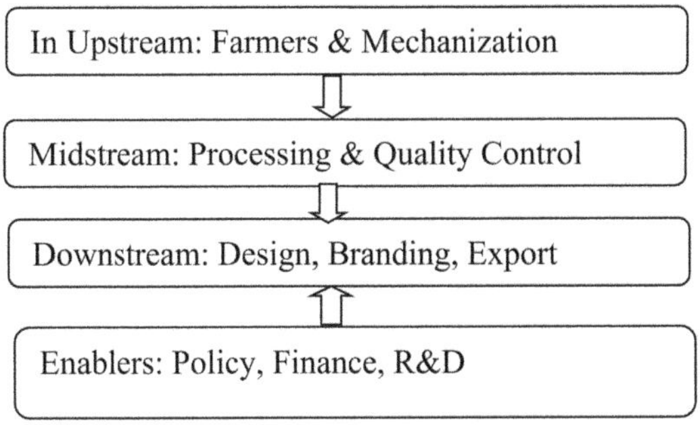

FIGURE 11.4 Integrated jute value-chain framework.

(Authors.)

This model moves beyond siloed interventions to a systemic, stakeholder-driven approach. For example, increasing land allocation for jute (upstream) must be matched with investment in backward linkages such as seed supply, equipment distribution, and training.

11.6 CONCLUSIONS

The jute industry in Bangladesh is under a severe sustainability crisis that is demonstrated by a shortage of labor, minimal mechanization, and financial isolation. The studies conducted on the major stakeholders reveal that although the mechanization of harvest might save labor time by 90%, it is incredibly expensive and lacks access to formal credit, as one-fifth of farmers have access to formal credit. This leads to 80% value addition being done outside the country, with the disparity between local and international retailers being large. The research suggests a National Integrated Jute Value-Chain Model to rejuvenate the so-called Golden Fiber to focus on government-assisted leasing of equipment, green financing, and the formation of design hubs to bridge the local artisans with global trends. This combined strategy will help generate three million jobs and transform Bangladesh into a sustainable leader for fiber. Although a small sample of qualitative research with a specific region is a limitation of the study, it lays the groundwork for further mixed-methods research of digital supply chains, the viability of green financing, and consumer preferences worldwide.

APPENDIX

See Table 11.6

TABLE 11.6
Respondents' Profile

No.	Code	Age	Gender	Education	Profession	Experience (Year)
1	BKBMGR	49	Male	Post-Graduation	Bank Manager	20
2	DAEOFF1	37	Male	Post-Graduation	Agriculture Officer	08
3	DAEOFF2	43	Male	Post-Graduation	Agriculture Officer	13
4	JTAGR1	27	Male	High School	Jute Farmer	5
5	JTAGR2	45	Male	Primary School	Jute Farmer	20
6	JTAGR3	37	Male	Primary School	Jute Farmer	15
7	JTAGR4	39	Male	Primary School	Jute Farmer	16
8	JTAGR5	42	Male	Primary School	Jute Farmer	20
9	BJRIOFF1	35	Male	Post-Graduation	Jute Research Officer	6
10	BJRIOFF2	45	Male	Post-Graduation	Jute Research Officer	15
11	BARIOFF3	34	Female	Post-Graduation	Jute Research Officer	5
12	JDPCOFF1	42	Male	Post-Graduation	Jute Promotion Officer	12
13	JDPCOFF2	37	Male	Post-Graduation	Jute Promotion Officer	7
14	JDPCOFF3	32	Female	Post-Graduation	Jute Promotion Officer	3
15	JDPCOFF4	45	Male	Post-Graduation	Jute Promotion Officer	16

REFERENCES

ADB. (2016). *Bangladesh Consolidating Export-Led Growth: Country diagnostic study*. Asian Development Bank (ADB). Malina. https://www.adb.org/publications/bangladesh-consolidating-export-led-growth

Ali, M. R., Kozan, O., Rahman, A., Islam, K. T., & Hossain, M. I. (2015). *Jute Retting Process: Present Practice and Problems in Bangladesh*. Kyoto University Research Information Repository (Kyoto University). http://hdl.handle.net/2433/201989

Asaduzzaman, M. (2022). *Jute Industry of Bangladesh: An Overview of Production and Export*. Emerging Credit Rating Ltd., Dhaka. https://emergingrating.com/wp-content/uploads/2022/10/Jute-Industry-of-Bangladesh_An-Overview-of-Production-and-Export.pdf

Azad, A. A. S., Ahmed, S., Oishi, Z. T., & Supta, S. S. S. (2020). An approach to design a sustainable supply chain: Future non-government investment in jute industry in Bangladesh. *Supply Chain Insider*, *5*(1), 34.

Barua, P., & Rahman, S. H. (2019). Sustainable value chain approach for vegetables product. *IUP Journal of Supply Chain Management*, *19*(1), 20–45.

Barua, P., Rahman, S. H., & Barua, M. (2021). Sustainable value chain approach for livestock-based livelihood strategies for communities of the southeastern coast of Bangladesh. *Modern Supply Chain Research and Applications*, *3*(3), 191–225. https://doi.org/10.1108/MSCRA-01-2019-0001

BBS. (2019). *Statistical Yearbook Bangladesh 2019*. Dhaka: Bangladesh Bureau of Statistics (BBS), Ministry of Planning.

BIDA. (2023). *Diversified Jute Products*. Dhaka: Bangladesh Investment Development Authority (BIDA).

Braun, V., & Clarke, V. (2006). Using thematic analysis in psychology. *Qualitative Research in Psychology*, *3*(2), 77–101. https://doi.org/10.1191/1478088706qp063oa

CEIC. (2023). *CEIC Data*. Retrieve on December 4, 2023. https://www.ceicdata.com

CSISA-MEA. (2022). *The Feed the Future Cereal Systems Initiative for South Asia Mechanization and Extension Activity Semi-Annual Report, 2021–2022*. Dhaka: Cereal Systems Initiative for South Asia – Mechanization and Extension Activity (CSISA-MEA).

DFID. (1999). *Sustainable Livelihoods Guidance Sheets*. London: DFID.

Gebresenbet, G., & Oodally, G. (2005). *Review and Analysis of Rural Agricultural Transport and Logistics in Developing Countries*. Technical Guidelines. Uppsala: Swedish University of Agricultural Sciences.

Guba, E. G. & Lincoln, Y. S. (1985). *Naturalistic Inquiry*. Thousand Oaks, CA: Sage.

Hossen, M. A. (2019). Mechanization in Bangladesh: Way of modernization in agriculture. *International Journal of Engineering Trends and Technology*, *67*(9), 69–77. https://doi.org/10.14445/22315381/IJETT-V67I9P212

Joshi, D., Panagiotou, A., & Rahman, M. W. (2024). *Agricultural Extension Services in Bangladesh: Experiences of Vulnerability, Access, and Benefits*. Colombo, Sri Lanka: International Water Management Institute (IWMI). https://hdl.handle.net/10568/159424

Kabir, M. & Islam, F. (2021). *Jute Trade: Bringing Back the Golden Glory of Jute Fibre in Bangladesh*. Dhaka: Sandhipath.

Kaplinsky, R., & Morris, M. (2000). *A Handbook for Value Chain Research* (Vol. 113). Brighton: University of Sussex, Institute of Development Studies.

Kirsch, M. S. (2018). *Agricultural Mechanisation & Rural Employment*. Ethiopia: Gesellschaft für Internationale Zusammenarbeit (GIZ).

Kumar, M., Bera, A., Gotyal, B. S., Naik, M. R., & Kumar, S. (2014). Technologies for sustainable jute fibre production. *Popular Kheti*, *2*(3), 47–52.

Moniruzzaman, S. M., Alam, M. J., Sharma, D., Begum, I. A., Patino, M. T. O., & McKenzie, A. M. (2024). Value chain analysis of jute fiber in Bangladesh. *Social Sciences & Humanities Open*, *10*, 101088. https://doi.org/10.3389/fsufs.2022.852178

New Age. (2020, June 30). Historic potential of jute sector ignored, yet again. https://www.newagebd.net/article/109821/historic-potential-of-jute-sector-ignored-yet-again

Petesch, P., & Caillava, I. R. (2012). *Voices of Young Villagers in Sub-Saharan Africa*. Background Paper. Washington, DC: World Bank.

Rifath, M. F. (2018). *A Conceptual Model of Upstream Jute and jute Diversified Products' Supply Chain Management*. Project Report. Dhaka: BRAC University.

Sadat, A., Ghosh, S. K., & Chakraborty, K. (2017). Impact of training on knowledge and adoption of jute technology in Uttar Dinajpur district of West Bengal, India. *Indian Research Journal of Extension Education*, *17*(2), 73–77.

Index

M

P

R

S

T

V

W

For Product Safety Concerns and Information please contact our EU representative GPSR@taylorandfrancis.com
Taylor & Francis Verlag GmbH, Kaufingerstraße 24, 80331 München, Germany

www.ingramcontent.com/pod-product-compliance
Ingram Content Group UK Ltd.
Pitfield, Milton Keynes, MK11 3LW, UK
UKHW022025061126
477531UK00002B/43
* 9 7 8 1 0 4 1 2 9 3 1 3 2 *